# DEVELOPMENTS IN SWEETENERS—2

# CONTENTS OF VOLUME 1

*Edited by* C. A. M. HOUGH, K. J. PARKER and A. J. VLITOS

# DEVELOPMENTS IN SWEETENERS—2

*Edited by*

## T. H. GRENBY,

B.Sc., Ph.D., C.Chem., F.R.S.C.

*Department of Oral Medicine and Pathology,
Guy's Hospital, London, UK*

and

## K. J. PARKER,

M.A., D.Phil., C.Chem., F.R.S.C.

## M. G. LINDLEY,

B.Sc., Ph.D., A.I.F.S.T.

*Tate & Lyle PLC Group Research and Development,
Philip Lyle Memorial Research Laboratory, The University,
Whiteknights, Reading, UK*

## APPLIED SCIENCE PUBLISHERS
LONDON and NEW YORK

05789994

APPLIED SCIENCE PUBLISHERS LTD
Ripple Road, Barking, Essex, England

*Sole Distributor in the USA and Canada*
ELSEVIER SCIENCE PUBLISHING CO., INC.
52 Vanderbilt Avenue, New York, NY 10017, USA


**British Library Cataloguing in Publication Data**

Developments in sweeteners—(The Developments series)
2
1. Non-nutritive sweeteners—Periodicals
I. Series
664'.5'005      TP422

ISBN 0-85334-202-4

WITH 17 TABLES AND 29 ILLUSTRATIONS

The selection and presentation of material and the opinions expressed in this
publication are the sole responsibility of the authors concerned

Printed in Great Britain by Galliard (Printers) Ltd, Great Yarmouth

D
664.5
DEV

# PREFACE

Sweet compounds almost universally induce a positive hedonic response in humans. It can be inferred from the reaction of the neonate to sugar that the liking for sweetness is inborn. It is an attractive concept, which has received much support, that the development of the sweet preference arose from man's need for instant energy. The ability to perceive bitterness, and the avoidance response so induced, may be associated with the extreme bitterness of many naturally occurring toxic compounds.

The per-capita consumption of nutritive sweeteners in the Western world is in excess of 100 lbs per annum, the major part of which is sucrose. Sucrose is not, of course, consumed simply because it is sweet. It has many functional properties that account for its widespread use in foods and beverages. For example, it is a bulking agent, it confers texture and mouthfeel, it acts as a preservative and it is fermentable.

The general public is being urged to control food intake to avoid obesity, the most obvious form of malnutrition in Western countries. The simplest way to reduce caloric intake is frequently perceived as a drive against the consumption of carbohydrate, particularly sugar. Thus there has been a growing demand by consumers on food manufacturers to develop reduced- or low-calorie foods. It perhaps would be more reasonable to urge consumers to reduce their level of intake of normal foods, rather than to require food manufacturers to create the illusion of normality. However, the public is also sensitive to the use of food 'additives' and substitutes, perceived as undue interference with the wholesomeness of the food supply, creating something of a dilemma for the industry.

The understandable desire to produce foods and beverages which are palatable can in many instances only be met in full by the use of nutritive sweeteners. The use of non-nutritive sweeteners inevitably raises problems in applications where the bulking agent is not water. The use of saccharin, in particular, results in products which are inferior from a flavour and textural standpoint.

For many decades saccharin was the only non-nutritive sweetener permitted for use in foods. With the discovery of cyclamate in 1937, an alternative became available which was found to provide a more acceptable quality of sweetness in admixture with saccharin. However, with the withdrawal of regulatory approval for cyclamate in 1969 in the USA and with many European countries following suit in 1970, food manufacturers in these countries were left with saccharin as the only permitted non-nutritive sweetener. Saccharin is generally considered to have a poor sweetness quality relative to sucrose as a consequence of its persistent bitter/metallic aftertaste. Thus the ban on cyclamate gave considerable impetus to the search for alternative high-intensity sweeteners.

The criteria by which non-nutritive sweeteners are judged include sweetness intensity and quality, safety in use, and cost of manufacture—the most important of these being quality of sweetness and freedom from any toxic effects, neither of which is predictable with any degree of confidence from a consideration of molecular structure. Not only is the quality of sweetness unpredictable but even the presence or absence of sweetness itself is only explicable retrospectively, and toxicity, or the lack of it, is equally unpredictable.

The sweetener issue is undoubtedly an emotive one, and it is perhaps because of the degree of interest the subject generates that so many developments have taken place in this area over the past few years. In compiling this volume in the Developments Series, we have tried to ensure that some significant recent advances are brought to the reader's attention. Novel bulk sweeteners, Lycasin®* and lactose hydrolysate syrups are covered in some detail from the physiology and production and applications standpoints. Considerable attention has also been given to the toxicology, safety evaluation and metabolism of some non-nutritive sweeteners. These reviews of clinical and metabolic developments, for both bulk and intense sweeteners, provide a necessary background for a full understanding of the dietary and nutritive role of both natural and artificial sweeteners. Recent developments in non-nutritive sweeteners and their practical applications in foods are also discussed. The only metabolic disorder that can be attributed incontrovertibly to dietary carbohydrate

---

* Lycasin® is a registered trade mark of Roquette Frères, France.

intake is dental caries, and it is appropriate that the evidence for this is presented in full.

Research into all aspects of sweeteners is multi-disciplinary, involving the synthetic chemist and the clinician, the psychologist and the food scientist. Because of this, it has not been possible to cover all aspects of sweeteners, but we hope that there is sufficient variety and interest in the contents of this volume to stimulate workers in the area towards even more new and exciting developments.

T. H. GRENBY
M. G. LINDLEY
K. J. PARKER

# CONTENTS

ix

# LIST OF CONTRIBUTORS

J. DANIEL

Life Science Research, Stock, Essex CM4 9PE, UK

T. H. GRENBY

Department of Oral Medicine and Pathology, Guy's Hospital, London SE1 9RT, UK

J. D. HIGGINBOTHAM

Tate & Lyle Group Research and Development, Philip Lyle Memorial Research Laboratory, The University, Whiteknights, PO Box 68, Reading, Berkshire RG6 2BX, UK

P. LEROY

Research and Development, Roquette Frères, 4 rue Patou, 62136 Lestrem, Pas de Calais, France

M. G. LINDLEY

Tate & Lyle Group Research and Development, Philip Lyle Memorial Research Laboratory, The University, Whiteknights, PO Box 68, Reading, Berkshire RG6 2BX, UK

I. S. MENZIES

Department of Chemical Pathology and Metabolic Disorders, St. Thomas's Hospital Medical School, London SE1 7EH, UK

A. G. RENWICK

> *Faculty of Medicine, Medical and Biological Sciences Building, University of Southampton, Bassett Crescent East, Southampton SO9 5TU, UK*

P. J. SICARD

> *Research and Development, Roquette Frères, 4 rue Patou, 62136 Lestrem, Pas de Calais, France*

D. SNODIN

> *Tate & Lyle Group Research and Development, Philip Lyle Memorial Research Laboratory, The University, Whiteknights, PO Box 68, Reading, Berkshire RG6 2BX, UK*

C. A. WILLIAMS

> *Department of Biochemistry, University of Surrey, Guildford, Surrey GU2 5XH, UK*

*Chapter* 1

# MANNITOL, SORBITOL AND LYCASIN: PROPERTIES AND FOOD APPLICATIONS

P. J. Sicard and P. Leroy

*Research & Development, Roquette Frères, Lestrem, France*

## SUMMARY

*This chapter describes the sources, methods of industrial production, and chemical and physical properties of the polyols, mannitol and sorbitol, and the hydrogenated glucose syrup Lycasin®,\* which was developed for its low cariogenic potential.*

*Food applications are discussed in detail, and recipes are given for 'sugar-free' confectionery based on mannitol, sorbitol and Lycasin. The types of confectionery include hard-boiled sweets, hard and soft coatings, chewy sweets, direct-compression tablets, and chewing gum, and it is explained how their characteristics can be improved by mixtures of the polyols.*

## 1. INTRODUCTION

For most of mankind carbohydrates are the principal source of calories. The main sources of nutritional carbohydrates are cereals, roots and tubers, the sugars sucrose and lactose, and fruit.

Cereals and tubers provide about 60% of the carbohydrates in human food.[1] Thus starch is the most widely consumed nutritional carbohydrate.

---

\* Lycasin® is a registered trade mark of Roquette Frères, France.

The other principal food carbohydrates, relatively few in number,[2] are:

1.  Sucrose, which comes second to starch as a source of carbohydrate;
2.  Lactose, present in milk; and
3.  Monomeric sugars, such as dextrose and fructose, which occur in the free state in fruit and honey.

The development of techniques for the industrial manufacture of food ingredients has resulted in the production, from naturally occurring materials, of carbohydrates with special properties. Examples are modified starches, pre-cooked starches, maltodextrins, glucose syrups, and invert sugar, a mixture of dextrose and fructose.

Polyols resulting from reduction of the reducing group of some of the above carbohydrates are also used in the food industry for special purposes. They consist principally of mannitol, found naturally as a constituent of the exudate of the manna ash (*Fraxinus rotundifolia*) and used in ancient times as a sweetener, sorbitol, which is found in the natural state in many fruits, and Lycasin® 80/55, a proprietary hydrogenated glucose syrup.

Although mannitol has for many years been obtained from natural sources by extraction, nowadays it is produced industrially by reduction of sugars, as are the other sugar alcohols. Reduction is normally carried out using hydrogen gas under pressure in the presence of Raney-nickel catalyst. In theory, about 125 litres of hydrogen at atmospheric pressure are required to manufacture 1 kg of a polyalcohol such as sorbitol, from glucose. Very strict safety precautions must be taken to avoid hazards, such as contact with atmospheric oxygen, since the catalysts are pyrophoric. The catalyst must be protected to avoid poisoning; failure to do so results in a loss of catalytic action and poor conversion on hydrogenation, with an adverse effect on both quality and productivity.

Sorbitol has been used for many years as an acceptable nutrient for diabetics because of its slow intestinal absorption and its non-insulin-dependent metabolism. Its uses in foods include that of humectant, and both sorbitol and mannitol are now being used to replace sucrose in the preparation of 'sugarless' products.

Lycasin 80/55 owes its origin to work done to avoid the cariogenic effects of sucrose. In response to the request of one of its confectionery customers,[3] the Lyckeby Starch Company proposed, about 15 years ago, that glucose syrups, raw materials widely used in confectionery manufacture, could be converted into their hydrogenated equivalents. The '-itols' were, in fact, known to be considerably less acidogenic than the

'-oses', from which they are derived, in the presence of *Streptococcus mutans* and other oral bacteria.[4]

Glucose syrups are obtained from starch by acid or enzymic hydrolysis, or a combination of the two. For each method used, the product has a well-established mono-, di- and higher-saccharide composition at a given degree of hydrolysis.[5] In principle, there can be as many varieties of hydrogenated glucose syrups as there are of glucose syrups. Lyckeby originally developed a range of Lycasin syrups based on potato starch hydrolysates with a dextrose equivalent (DE, a measure of the degree of hydrolysis) between 15 and 75. The type used for 'Lycasin hard candy' confectionery was derived from a 25 DE base, the types derived from higher DE bases having a high sorbitol content since increasing the extent of acid hydrolysis liberates an increasing proportion of dextrose.

This range was subsequently developed by Lyckeby (Lycasin) in conjunction with Roquette S.A. (Polysorb®*), the latter company being specialists in cornstarch hydrolysis. By agreement, Lycasin has for some years been the exclusive property of Roquette S.A., and Lycasin 80/55 represents the most fully developed of the hydrogenated glucose syrups. It has been specially developed so as not to cause a severe drop in pH in dental telemetry tests, and consequently it can be considered as 'tooth-preserving', whilst at the same time having the technological qualities necessary for the manufacture of sweets of the hard-boiled type.

## 2. MANNITOL

Mannitol derives its name from the fact that it is the main constituent of manna, a sweet exudate of the flowering ash (*Fraxinus ornus*) from which it was isolated by Proust.[6] D-Mannitol is a hexitol widely found in nature in higher plants (ash, olive, fig and celery), in fungi (*Lactarius* and *Agaricus* may contain up to 15–20% of dry extract as mannitol) and in numerous algae, in particular brown seaweeds such as *Laminaria* spp. which can contain up to 20% of mannitol according to the season.

### 2.1. Production and Properties
Mannitol was, for a long time, a product of limited availability since it was initially obtained by alcohol extraction from natural sources, namely manna (from *Fraxinus rotundifolia*) and seaweed.[7] It has also been

* Polysorb® is a registered trade mark of Roquette Frères, France.

obtained as a by-product of certain fermentations using *Aspergillus*,[8,9] *Torulopsis*[10] and *Candida*.[11] Certain strains of *Lactobacillus*[12,13] have also been quoted as a source of mannitol by fermentation, but these processes do not seem to have been used in practice.

In terms of chemical structure, mannitol is the hexitol corresponding to the aldohexose mannose. However, since this carbohydrate is not available industrially, the traditional raw material for obtaining mannitol is sucrose which, after hydrolysis, can be hydrogenated to produce a mixture of sorbitol and mannitol. The reduction of dextrose in an alkaline medium has also been used, the alkalinity resulting in epimerisation of dextrose to fructose, i.e. a situation analogous to the previous example. Mannitol is isolated by fractional crystallisation, taking advantage of its lower solubility in water compared with sorbitol. Recrystallisation is necessary to obtain high-purity mannitol. Part of the mannitol inevitably remains in solution and the cost of the recrystallisation operation adds to the cost of the product. Theoretically, mannitol and sorbitol are formed from sucrose in a 1:3 ratio (Fig. 1). In practice, the yield of mannitol is slightly increased by reduction in an alkaline medium,[14,15] although the yield after crystallisation does not exceed 15 to 18 %.

In order to reduce the cost of mannitol produced from dextrose, processes have been sought which would make it possible to obtain a higher mannitol content after reduction; the enzymic isomerisation of dextrose using glucose isomerase provides, at equilibrium, a dextrose–fructose

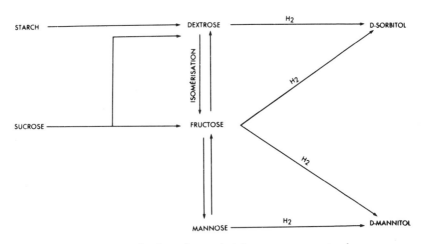

FIG. 1.    Production of mannitol from sucrose or starch.

## TABLE 1
### PROPERTIES OF MANNITOL

1. Chemical names and synonyms    D-mannitol, mannite, manna sugar
2. CAS registry number    87-78-5 and 69-65-8
3. Formula and structure    $C_6H_{14}O_6$

$$\underset{\displaystyle \overset{\displaystyle |}{OH}\quad \overset{\displaystyle |}{OH}}{HOCH_2-CH-CH-\overset{\displaystyle \overset{OH}{|}}{CH}-\overset{\displaystyle \overset{OH}{|}}{CH}-CH_2OH}$$

4. Molecular weight    182·17
5. Solubility in water

| | | | | *Temperature* (°C) | | | | |
|---|---|---|---|---|---|---|---|---|
| | | *0* | *10* | *20* | *30* | *40* | *50* | *60* |
| g 100 g$^{-1}$ water | | 10 | 14 | 17 | 25 | 34 | 45 | — |
| Concentration of saturated solution (% w/w) | | 9·1 | 12·3 | 14·5 | 20 | 25 | 31 | 37·5 |

6. Sweetness[17] (sucrose = 100)    57
7. Melting point    165–168 °C
8. Heat of solution    $-29$ cal g$^{-1}$

composition comparable with that obtained by hydrolysis of sucrose. A process for the catalytic epimerisation of glucose into mannose was described in 1975.[16]

The properties of mannitol are shown in Table 1. The existence of special crystalline forms of mannitol has been indicated.[18,19]

### 2.2. Applications and Methods of Use

The importance of manna in providing food for the people of Israel during the exodus from Egypt is reported in the Bible. Manna was used for centuries in Mediterranean areas for cake making and to sweeten medicines for children. The technological reasons for using mannitol are its pleasant taste, high stability and non-hygroscopicity. Like other polyols with an agreeable taste, it is currently used for preparing 'sugar-free' products. Its low solubility, however, prevents it from being used in products such as ice-cream, canned fruits, soft drinks or confectionery. Mannitol is principally used in the manufacture of:

1.  'Sugarless' chewing gum, particularly for dusting slabs during rolling and cutting-out operations, as an anti-adhesion agent. It can also be added to the mix, but in small quantities (less than 5 %) as an agent inhibiting the crystallisation of the other polyalcohols used in the formulation; and

2.  'Tablets'—since the industrial types available are fine powders, mannitol must be granulated prior to being compressed.

## 3. SORBITOL

Sorbitol was given its name by the French chemist Boussingault,[20] who discovered it in 1872 in the berries of the mountain ash (*Sorbus aucuparia* L.). Although it is found widely in plants and fruits,[21] sorbitol has been produced industrially for many years by the reduction of dextrose.

### 3.1. Production and Properties

Sorbitol was first synthesised industrially some 50 years ago and subsequent improvements to the processes and the raw materials used have enabled it to be produced on an economically acceptable basis. While electrochemical reduction[22,23] was used initially, the standard production method is now catalytic hydrogenation under pressure. Two raw materials may be used: sucrose or starch.

Using sucrose, which is hydrolysed to invert sugar before hydrogenation, a mixture of the two stereoisomeric hexitols sorbitol and mannitol is formed. Since mannitol is much less soluble in water than sorbitol, it is mainly recovered by fractional crystallisation. A proportion of the mannitol nevertheless remains in solution; this explains the tolerant attitude of the US Pharmacopoeia (USP) and the Food Chemicals Codex which specify a minimum of only 91 % D-sorbitol.

When starch is used, it must first be hydrolysed. Enzymic hydrolysis processes have made it possible to obtain very pure dextrose. Subsequent electrochemical reduction of dextrose in an alkaline medium leads to the formation of mannitol in considerable quantities as a result of the alkali-catalysed epimerisation of dextrose to levulose (fructose), a ketohexose which is reduced more rapidly than dextrose.

With the catalytic hydrogenation of dextrose, on the other hand, it is possible to produce sorbitol containing less than 2 % mannitol. Dextrose prepared from starch, nevertheless, contains small amounts of disaccharides after crystallisation, and, for the USP quality of sorbitol, 1 % total sugars (reducing sugars after hydrolysis) is acceptable.

Sorbitol solutions containing higher levels of total sugars and complying with the E420-(ii) criteria[24] are prepared from starch hydrolysates which are very rich in dextrose. After hydrogenation, the syrups are purified and demineralised on ion-exchange resins (for removal of gluconic acid) and then further purified and concentrated to 70 % w/w. Solid sorbitol is obtained from the purest solutions by evaporation and crystallisation, a technique which normally produces amorphous sorbitol with a low melting point (92 °C).[25] Under special conditions,[26] however, solid sorbitol is obtained with a stable crystalline structure ($\gamma$-sorbitol) and with a melting point above 96 °C (Table 2).

The high negative heat of solution of sorbitol combined with its high solubility in water, explain the pronounced 'cooling' or 'fresh' taste of solid

TABLE 2

PROPERTIES OF SORBITOL

| 1. Chemical names and synonyms | D-sorbitol, D-glucitol, sorbite |
| 2. CAS Registry number | 50-70-4 |
| 3. Formula and structure | $C_6H_{14}O_6$ |

$$HOCH_2-\underset{\underset{OH}{|}}{CH}-\underset{\underset{OH}{|}}{CH}-\underset{\underset{OH}{|}}{CH}-\underset{\underset{OH}{|}}{CH}-CH_2OH$$

| 4. Molecular weight | 182·17 |
| 5. Solubility in water | |

| | *Temperature* (°C) | | | | | |
| | 0 | 10 | 20 | 30 | 40 | 50 |
|---|---|---|---|---|---|---|
| g 100 g$^{-1}$ water | 147 | 180 | 220 | 270 | 360 | 500 |
| Concentration of saturated solution (% w/w) | | 59·5 | 64·3 | 68·7 | 73 | 78·3 | 83·3 |

| 6. Sweetness[27] (sucrose = 100) | 50–60 |
| 7. Melting point[28] | Metastable form—93 °C |
| | Stable ($\gamma$-) form—96 °C |
| | Anhydrous form—110–112 °C |
| 8. Heat of solution | −28 cal g$^{-1}$ |

sorbitol. The formation of a saturated solution of sorbitol in 100 g water requires the absorption of slightly over 9000 cal (cf. 984 cal for sucrose). Sorbitol is only slightly soluble in ethyl alcohol and is practically insoluble in other organic solvents. The solubility of sorbitol in aqueous alcohol decreases with increasing alcohol concentration. An aqueous solution of ethyl alcohol at 10° Gay Lussac (GL) contains, when saturated at a temperature of 20 °C, 66 % sorbitol and 34 % alcohol (Fig. 2). On the other hand, ethyl alcohol at 90° GL is only capable of dissolving 3 % sorbitol at 20 °C, while alcohol at 40° GL contains 55 % sorbitol at saturation.

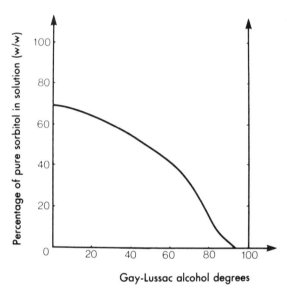

Gay-Lussac alcohol degrees

FIG. 2.   Solubility of sorbitol in water–ethyl alcohol mixtures at 20 °C.

Sorbitol is miscible in all proportions with polyhydric alcohols such as glycerol and glycols. This is of prime importance for some applications, for example in cosmetics and tobacco.

The limiting relative humidity and the stable crystalline form is 70–73 % at 20 °C (cf. 80–84 % for sucrose and 97 % for mannitol). Crystalline sorbitol in its stable form is, however, only slightly hygroscopic.

Sorbitol solutions, as supplied commercially, are viscous colourless odourless syrups with a moderately sweet taste.

As already stated, sorbitol is very soluble in water and the usual concentration of commercial solutions is $70\%$ w/w. This is above the saturation solubility of pure sorbitol at $20\,^\circ\mathrm{C}$. There is a tendency, therefore, for crystallisation to occur in solutions of pure sorbitol when the temperature falls below about $23\,^\circ\mathrm{C}$, although the solutions will often remain in a super-saturated state. To reduce the tendency to crystallise, so-called 'non-crystallising' grades have been developed which contain hydrogenated oligosaccharides in addition to sorbitol.

Viscosity increases with concentration (Fig. 3) but decreases steeply with an increase in temperature. At equal concentrations, solutions of sorbitol are more viscous than those of polyhydric alcohols with a lower molecular weight.

The ability of sorbitol to act as a humectant has resulted in solid sorbitol being regarded as hygroscopic. The hygroscopicity of sorbitol powder is, however, a function of its crystalline form, and is very small. Hence humectancy is a property of sorbitol solutions, not of the solid.

The equilibrium relative humidity (ERH) of aqueous solutions is lower

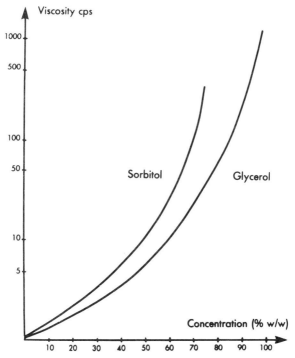

FIG. 3. Viscosity of solutions of sorbitol and glycerol.

for glycerol than for sorbitol (Fig. 4). Sorbitol solutions have, on the other hand, the advantage of a high resistance to change in water content. In a humid atmosphere a solution of sorbitol takes up less water than glycerol and, in a dry atmosphere, loses absorbed water less rapidly. Its capacity to stabilise relative humidity is superior to that of glycerol (Fig. 5). This results in a greater weight stability and smaller physical changes in products containing sorbitol.

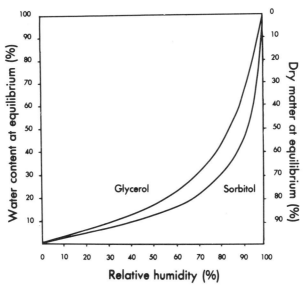

FIG. 4.   Water content at equilibrium of aqueous solutions of sorbitol and glycerol as a function of relative humidity.

Sorbitol is very stable and chemically unreactive. The absence of reducing capacity explains the fact that practically none of the well-known sugar reactions occur with sorbitol. Thus it does not reduce Fehling's solution and does not undergo a Maillard reaction in the presence of amino compounds. Sorbitol is very stable to heat, withstanding high temperatures without foaming or discoloration, and it can therefore be added to food products before they are cooked.

### 3.2. Applications
The food applications of sorbitol are based on the use of its specific properties, such as viscosity and humectancy. It combines moderate

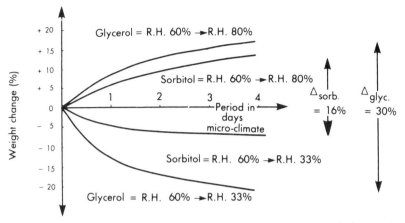

FIG. 5.   Capacity of aqueous solutions of sorbitol and glycerol to stabilise relative humidity.

sweetening power with special taste characteristics such as its 'cooling' effect. Sorbitol combines well with other food raw materials such as sugars, gelling agents, albumin and vegetable fats.

The properties utilised will differ according to whether sorbitol constitutes a small proportion of the carbohydrate of the food product or whether it is used in large amounts to replace sugar completely. In the first case, the advantage of sorbitol is based on its special properties as an additive, that is, as a humectant, texturising and anti-crystallising agent, and its ability to chelate heavy metals. In the second case, its value lies in its metabolic properties as a complete replacement for sucrose in sugar-free products such as jams, chewing gum, tablets and coatings. Sorbitol can only be used for this type of product insofar as it possesses the required physical properties.

### 3.2.1. Low-Level Uses of Sorbitol
The moisture-stabilising and textural properties of sorbitol are used in confectionery, biscuit manufacture, and cake and chocolate manufacture where products have a tendency to become dry or to harden during storage. Sorbitol improves the texture of the products and acts as an anti-crystallising agent towards soluble substances in the liquid phase.

*3.2.1.1. Sorbitol and confectionery.* Although the main carbohydrate raw materials used in confectionery are sucrose and glucose syrup (corn syrup), sorbitol has the advantage of being chemically inert—it will not cause inversion when used in mixtures with sucrose, even at a high temperature. Invert sugar is rightly regarded as an excellent softening agent, but it nevertheless has the disadvantages that products containing it are very sensitive to changes in atmospheric moisture, leading to 'cold-flow' in confectionery and desiccation in biscuits. Sorbitol lessens these effects of invert sugar by virtue of its moisture-stabilising qualities.

In admixture with other sugars, sorbitol modifies crystalline structure and improves smoothness, colour and shelf-life. Thus fondants and creamy sweets containing sorbitol have finer-grained, more homogeneous, whiter and more stable textures. Sorbitol acts as a partial solvent for colouring materials and, because of sorbitol's chemical inertness, stability is increased and diffusion throughout mixtures is assisted.

Sorbitol is not suited for the production of boiled sweets and dry, brittle products, such as hard nougats. These have a very low ERH and tend to take up water from the atmosphere; it is therefore important to use constituents of low hygroscopicity in their manufacture. In the case of centre-filled high-boiled candies, on the other hand, the addition of small quantities of sorbitol (1–2 %) gives the shell of the sweets better plasticity, making mechanical processing easier. The liquid centre of such candies must have a solids content sufficient to prevent the outer case redissolving during storage, and must also contain sufficient material to prevent sucrose from crystallising out. Sorbitol will act as such an anti-crystallising agent and will also allow the viscosity of the liquid centre to be lowered without reducing the solids content.

With products such as soft caramels, soft nougats and chewy sweets the addition of a small percentage of sorbitol allows smooth and creamy products of good texture to be made and reduces adhesion to teeth during chewing. The relative reduction in viscosity, if sorbitol is introduced in considerable quantities, can result in the appearance of sucrose microcrystals.

The use of sorbitol is particularly appropriate in fondants. In deposited cream fondants it can be introduced either during initial cooking of the sucrose–corn syrup mixture or as 'bob' syrup at the depositing stage. The main objectives in adding sorbitol are to obtain:

1.   An increase in the liquid phase and its dry matter content, and a decrease in the ERH of the liquid phase;
2.   Control of crystal dimensions;
3.   Softening of the fondant produced;

4. Whiteness of the fondant (sorbitol does not yellow on cooking); and
5. A reduction in threading at the depositing stage.

The humectant properties of sorbitol account for its use in jellies and gums. These products have a rather high ERH and tend to dry-up and harden when stored under normal ambient conditions; sorbitol improves their stability in this respect.

In the case of almond pastes, sorbitol makes them more supple and normally delays the onset of rancidity by chelating with the metals which catalyse this process in the presence of air.

With candied fruits, the fluidity of sorbitol enables the viscosity of the candying solutions to be lowered without reducing the solids content. It also acts as an anti-crystallising agent towards sugar and furthermore allows less sticky and lighter-coloured products to be manufactured as there is no colouring during heating. Candied fruits are more supple, and this reduces losses during manufacture through crumbling and breakage of the fruit. The reduction in the viscosity of the candying syrup makes it easier to remove the surface layer of syrup after candying and improves the resistance of the sucrose glazing during subsequent operations.

*3.2.1.2. Sorbitol in biscuit and cake making.* Sorbitol has a plasticising effect which improves the texture of these products, and its moisture-stabilising action protects them from drying and maintains their initial freshness during storage.

Sorbitol forms colour less readily than the sugars, which confirms its advantage in biscuit manufacture in the production of yellow doughs where a fresh colour is particularly sought after.

*3.2.1.3. Sorbitol as an anti-freeze.* Sorbitol syrups are used as an alternative to glycerol in spoonable, soft-scoop ice-cream formulations. These syrups, at a level of 3–5%, enable ices to be softened at a low temperature as a result of the increase in the liquid phase.

'Fish minces' are normally used as a protein base in the manufacture of products such as kamaboko, or in fish delicatessen products since they have a flavour which is more or less neutral depending on the extent of washing. These fish pastes are preserved before use by freezing. Their various applications require the maintenance of the physico-chemical stability of the proteins during this storage in a deep-frozen state in order to preserve the initial functional properties and colour of the fish fibres. Sorbitol allows these properties to be conserved; it is an excellent freeze protector when used at a level below 4 % in the presence of polyphosphates.

### 3.2.2. High-Level Uses of Sorbitol in 'Sugar-Free' Products

In the applications described above, sorbitol is only used at low levels as an additive with special technological properties. However, sugar-free products with a high sorbitol content are produced in which sorbitol is totally substituted for sucrose.

### 3.2.2.1. Sorbitol confectionery.

It is possible to make boiled sweets using sorbitol, but only by boiling the mix under vacuum at a high temperature until the residual moisture content is below 1 % and then depositing it at 70–80 °C. Cooling must be completed in a dry atmosphere (ERH below 30 %) because the product is hygroscopic. Special packaging has to be used to ensure adequate protection against moisture uptake.

### 3.2.2.2. Directly-compressed sorbitol powder.

Sorbitol powder, with the addition of various ingredients, such as lubricants and flavours, can be compressed directly into relatively hard tablets. Grades with special particle sizes have been developed for the manufacture of hard tablets for sucking and soft tablets for chewing.

The suitability of sorbitol for compression depends on its particular physical characteristics, which are, in turn, governed by the crystallisation method used. Thus each manufacturer of sorbitol supplies powder with specific properties. The most important of these for direct compression uses are:

1.    The shape of the particles, which must flow freely;
2.    A low tendency to disintegrate so that continuous handling and transport can be effected without damage to the particles;
3.    A stable crystalline structure to avoid lumping of the powders and to reduce hygroscopicity; and
4.    A special surface state, essentially dendritic, to allow strong interparticle cohesion.

### 3.2.2.3. Coating with sorbitol.

Coating of sweets and other products can be effected with sorbitol. The solubility characteristics of sorbitol and the viscosity characteristics of its saturated solutions make it necessary to operate under different conditions from those normally employed in sugar-coating using sucrose.

For the production of soft coatings a non-crystallisable liquid film of hydrogenated glucose syrup is deposited on the centres and sorbitol powder is scattered on this film. Successive additions of non-crystallisable syrup are made, followed by sorbitol powder, until the desired thickness of coating is obtained.

In the production of hard coatings the crystallising sorbitol syrups employed need to have the following physical properties in order to obtain a regular and homogeneous crystallisation from the layer of syrup:

1.  The viscosity must be sufficiently low to allow the liquid film to be distributed evenly; and
2.  The solubility must be such that the syrup crystallises rapidly under the action of warm air introduced into the pan.

The temperature of the sorbitol syrup at 70% solids must not exceed 30–35 °C, with the centres maintained at this temperature, whereas that of a sucrose syrup must be higher in order to increase the crystallisation yield (85 °C at 80% solids), with the centres maintained at a minimum of 50 °C.

Coating with sorbitol is readily effected without the need to concentrate the commercial crystallising syrup (70% solids), or to heat the centres excessively. Thus heat-sensitive products, such as chewing gum and chocolate, can be coated.

*3.2.2.4. Sorbitol–Lycasin 80/55–mannitol fondants.* Traditional fondant is made by beating a supersaturated sucrose solution. A 'sugar-free' fondant can be produced with a mixture of sorbitol, Lycasin 80/55 and mannitol, in a ratio of 50:40:10, respectively, boiled at 118–122 °C, to give the consistency desired. After cooling to 60 °C, the boiled syrup is beaten. The fondant obtained is very white, finely crystallised and fresh-tasting.

*3.2.2.5. Chewing gum.* Sucrose can be replaced by sorbitol in the manufacture of chewing gum, either as a solution in the liquid phase or as a powder in the solid sweetening phase. If required, glycerol may be added as a plasticising agent and mannitol as an anti-sticking agent.

The plastic properties of chewing gum may be improved by adding Lycasin 80/55. A chewing-gum formulation must meet two requirements:

1.  It must allow the supple texture of the product to be maintained; and
2.  The ease of machine working must be acceptable.

The preservation of a supple texture is controlled by the choice of the liquid phase of the chewing gum, which acts as a crystallisation inhibitor towards sorbitol and other soluble components under changes in atmospheric humidity. The formulation of a suitable sorbitol-based chewing gum relies principally on the nature of the anti-crystallising liquid phase, which is basically a non-crystallising sorbitol syrup to which is added soluble components, such as gluconates, mannitol, glycerol and gum arabic, which inhibit the crystallisation of sorbitol.

An example of a formulation for sorbitol chewing gum is:

| | |
|---|---|
| Dreyfuss base 34/42 | 30 % |
| Sorbitol powder, Neosorb® * 60 | 53 % |
| Sorbitol solution, Neosorb 70/70 | 14 % |
| Glycerol | 2 % |
| Flavouring | 1 % |

It should be noted, however, that the ideal crystallisation inhibitor for sorbitol is the hydrogenated glucose syrup Lycasin 80/55 rather than a non-crystallising sorbitol solution.

*3.2.2.6. Jams and preserved fruit in syrup.* Since sorbitol gives a high fluidity to gels, syneresis can occur so that pectins with a low methoxy level must be used as gelling agents. Jellies and jams using sorbitol are clear, bringing out the colour of the fruit without masking the flavour.

The manufacture of preserved fruit in syrup does not present any special problems using sorbitol.

*3.2.2.7. Sorbitol in chocolate manufacture.* Sucrose can be replaced completely by sorbitol in the manufacture of chocolate, although the characteristics of sorbitol make it necessary to work at a low temperature in order to avoid granulation by adhesion of the crystals during conching. To reduce the load on the cylinders, it is preferable to use a fine-grain stable crystalline form of sorbitol, although too fine a particle size, and hence too large a crystalline surface, causes undue fixation of fats, giving a dry appearance to the finished product. To avoid processing difficulties, it is extremely important that all the ingredients should be very dry, owing to the tendency of microcrystals to reassociate after milling and during conching if moisture is present. Thus the milk powder, for example, should have a moisture content below 3 %.

---

* Neosorb® is a registered trade mark of Roquette Frères, France.

*3.2.2.8. Sorbitol lozenges.* Lozenges are normally produced from a base paste composed of very fine sugar bound together with gums. This paste is rolled, punched out and dried. Sorbitol lozenges can be made using the same technique, a typical formula for which is:

Binding agent

| | |
|---|---|
| water | 67 kg |
| sorbitol solution (70 % solids) | 45 kg |
| Bloom 240 gelatin | 6 kg |
| gum arabic | 8 kg |
| glycerol | 6 kg |

Sorbitol lozenges

| | |
|---|---|
| binding agent | 12 kg |
| very fine sorbitol powder ( < 100 $\mu$m diameter) | 79 kg |
| flavouring | |

*3.2.2.9. Miscellaneous.* When establishing a formulation for 'sugar-free' dietetic products, sorbitol can also replace sucrose in food products such as ice-creams (with a slight lowering of the freezing point of the mix), beers with a reduced alcohol content, liqueurs, and cream fillings for wafer biscuits.

In addition to its applications in the food and dietetic fields, sorbitol has a large number of other uses based on its special properties—the fields covered include the pharmaceutical, cosmetic, chemical (as in the production of vitamin C and polyurethane resins) and tobacco industries.

# 4. LYCASIN 80/55

Lycasin 80/55 is a hydrogenated glucose syrup with a particular defined composition.[29] It was developed primarily to meet special technological demands in the production of hard-boiled sweets and the need, recognised in countries such as Switzerland, for a low-acidogenic, and thus low-cariogenic, sweetener.[30]

## 4.1. Production

The hydrolysate obtained from the enzymic hydrolysis of corn starch is purified by conventional procedures used for the production of corn syrups, i.e. filtration on a pre-coated filter, decolorisation and de-mineralisation by ion-exchange resins. The reducing aldehyde groups of the terminal glucose units of the component saccharides are then reduced by hydrogen under pressure in the presence of Raney–nickel as catalyst. After hydrogenation, the syrup is further purified and concentrated to the solids content required. Although it is technically possible to obtain a dried powder from this syrup, the hygroscopicity of the product necessitates special conditions for production and storage which are not commercially economic.

### 4.2. Composition of Lycasin 80/55

Lycasin 80/55 is a hydrogenated derivative of partly hydrolysed starch, comprising of the hydrogenated homologues of dextrose and the dextrose oligomers present in a specific corn syrup obtained by enzymic hydrolysis. Total hydrolysis of the hydrogenated saccharides liberates sorbitol and dextrose. The overall approximate composition of Lycasin 80/55 is:

| | |
|---|---|
| Sorbitol | 44 % |
| Dextrose | 56 % |

In fact, there is no free dextrose in Lycasin 80/55 and the level of free reducing sugars is less than 0·2 %. This means that hydrogenation of the terminal aldehyde groups is virtually complete. The free sorbitol content is 6–8 % since there is always some dextrose present in the syrup from the hydrolysis of starch. It is important to use hydrolysis conditions such that the quantities of dextrose liberated are minimal, so as to have the smallest possible quantity of free sorbitol in the hydrogenated product.

The typical composition of Lycasin 80/55, measured by gel permeation chromatography or by high performance liquid chromatography, is shown in Table 3.

### 4.3. Properties

Lycasin 80/55 is a clear colourless odourless syrup with an agreeable sweet taste. Its general specification is given in Table 4.

Lycasin 80/55 does not crystallise even at low temperature or high concentration, and like glucose syrups, inhibits crystallisation of other components present in confectionery recipes.

The viscosity of Lycasin 80/55 at 65 and 75 % solids is higher than that of sucrose at 65 % concentration, and decreases steeply with increases in temperature (Fig. 6).

The commercial syrup is only slightly hygroscopic. On the other hand,

TABLE 3
TYPICAL ANALYSIS OF THE COMPOSITION OF DRY LYCASIN
80/55

| Component | Percentage present |
|---|---|
| DP 1 (Sorbitol) | 7·0 |
| DP 2 (Maltitol) | 52·5 |
| DP 3 | 17·5 |
| DP 4 | 1·0 |
| DP 5 | 2·0 |
| DP 6 | 2·5 |
| DP 7 | 3·5 |
| DP 8 | 2·2 |
| DP 9 | 1·0 |
| DP 10 | 1·0 |
| Between DP 11 and DP 20 | 9·0 |
| >DP 20 | 0·8 |

Lycasin 80/55 in the solid state is very hygroscopic with an ERH below 20%. Precautions must therefore, be taken over the packaging of some confectionery products.

Lycasin 80/55 has a very agreeable sweet taste—its sweetness being 0·75 times that of sucrose on equal weight basis. This is sufficient for confectionery use without the need to add high-intensity sweeteners. It should be noted that because the sweetness is less than that of sucrose, the taste of food acids currently employed, such as citric and lactic acids, is more

TABLE 4
LYCASIN 80/55 SPECIFICATION

| | |
|---|---|
| Dry substance | $75\% \pm 1$ |
| Refractive index at 20 °C | 1·4762–1·4815 |
| Specific gravity at 20 °C | 1·36 |
| Reducing sugars | $0·2\%$ maximum |
| Specific rotation $\alpha_D^{20}$ | $+115°–120°$ |
| Viscosity (75% solids) | 2000 cP |
| pH in solution | 5–7 |
| Resistivity | 500 000 ohm-cm min |
| Sulphated ash | $0·1\%$ maximum |
| Sulphate | $100 \, \text{mg kg}^{-1}$ maximum |
| Chloride | $50 \, \text{mg kg}^{-1}$ maximum |
| Heavy metals | $10 \, \text{mg kg}^{-1}$ maximum |
| Nickel | $1 \, \text{mg kg}^{-1}$ maximum |

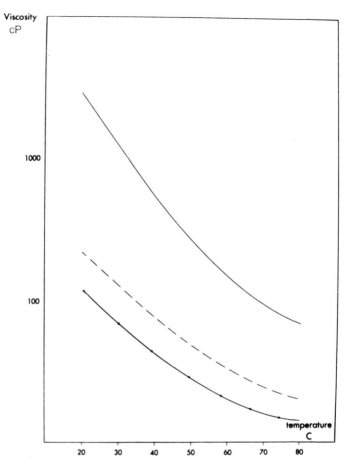

FIG. 6.  Viscosity as a function of temperature for Lycasin 80/55 and sucrose. ——,
Lycasin 80/55 (75% solids); – – –, Lycasin 80/55 (65% solids); ·——·, sucrose
(65% solids).

readily detected. It is therefore desirable to reduce the quantities of
acidulants normally employed in confectionery by 25 to 30%.

As in the case of sorbitol, the absence of free reducing groups means that
Lycasin 80/55 is very stable and inert chemically. With Lycasin 80/55 it is
possible to manufacture sweets at high temperature which are bright and
colourless and which are not subject to inversion by the acidic ingredients
usually used.

## 4.4. Applications

The use of Lycasin 80/55 for the preparation of products with a high water content, such as jams, fruit preserves, syrups and soft drinks, obviously does not pose any particular technical problem.

Certain properties of Lycasin 80/55 allow it to be used in applications where other sucrose substitutes are not suitable. This is the case with certain confectionery products.

### 4.4.1. Hard-Boiled Candies

The original reason for developing Lycasin 80/55 was to enable boiled sweets to be prepared without the use of sucrose or glucose syrup. The manufacture of such a candy with Lycasin 80/55 is simple. It is sufficient to dehydrate the Lycasin 80/55 *as completely as possible* and then to add flavouring. Because the addition of water must be avoided, it is necessary to use non-aqueous flavours.

High temperature and a vacuum are required for the boiling of Lycasin 80/55 in order to obtain the classical vitrified structure of boiled sugar. The residual water content must be less than 0·5–1 % instead of the usual 2–3 % in a traditional boiled sweet, otherwise the product very quickly becomes semi-plastic. The candies obtained are very clear and bright. They are hygroscopic, especially as surface crystallisation cannot occur if moisture is absorbed. It is, therefore, necessary to use a particularly moisture-proof packaging in order to prevent the sweets from becoming sticky. Packaging must be carried out very quickly, immediately following manufacture. With thin-film boiling lines where the boiled syrup is deposited in moulds on a cooling belt, boiling at 175 °C with a vacuum of 200–300 mm Hg results in a residual water content of the order of 0·5 %. After addition of flavours (free from water) and acidifying (acid level reduced by 25–30 %), it is best to deposit smaller-sized candies in order to hasten cooling and removal from the mould, given that Lycasin 80/55 based candies remain plastic for a longer time. On conventional production lines (boiling, cold table, mixer, roller, press) boiling must be carried out under conditions of particularly high temperature and vacuum, for example 160 °C and 720 mm Hg minimum. These conditions do not cause yellowing of the Lycasin 80/55, and allow a residual water content of less than 1 % to be reached in the boiled syrup. This, because of its temperature, is very liquid and must be cooled on a cold table in order to reach the plastic consistency necessary for working in the mixer. The flavours and acidulant are incorporated at this stage, at a temperature below that used in traditional confectionery processes, since the plasticity of cooked Lycasin is greater at a given temperature. After passing through the roller and press, packaging of the sweets must be carried out as quickly as possible.

### 4.4.2. Chewing Gum
Conventionally, a chewing gum comprises three phases:

| | | |
|---|---|---|
| 1. | An insoluble phase of gum base | 20–30% |
| 2. | A crystalline phase (sucrose in traditional formulations) | 50–60% |
| 3. | A liquid phase comprising a glucose syrup (85% solids) | 20% |

All the properties of the gum, namely, appearance, texture, chewability, stability without hardening, and machinability, are dependent on the equilibrium between these three phases, in which the liquid phase plays a vital part. It is important that the ERH of this liquid phase is of the order of 60–70%, these being average values for atmospheric humidities in Europe.

In 'sugarless' chewing gums, the sucrose is replaced by crystalline sorbitol, while the liquid phase normally consists of sorbitol syrup. The solid component of the liquid phase is then the same as the powder phase, so that it is necessary to add anti-crystallising agents such as glycerol. The liquid phase, nevertheless, remains at a high ERH, so that the formulations are insufficiently stabilised. Water exchange with the atmosphere is accompanied by a loss of plasticity with hardening and recrystallisation. Lycasin 80/55 reaches equilibrium at about 85% solids in an atmosphere of 66% relative humidity (Fig. 7). If Lycasin 80/55 is concentrated to 85% solids, a liquid phase is obtained which posseses excellent plasticising properties for the manufacture of sugarless chewing gum. This concentration of Lycasin 80/55 can be achieved by boiling the commercial syrup under atmospheric pressure at 113–114 °C.

### 4.4.3. Chewy Sweets
Lycasin 80/55 can be used in the manufacture of other confectionery items, such as chewy sweets. A typical recipe is:

| | |
|---|---|
| Lycasin 80/55 | 110 kg |
| Fats (melting point 34–36 °C) | 9 kg |
| Lecithin | 0·05 kg |
| Gelatin (soaked in 3 litres of water) | 1·3 kg |
| Citric acid | 0·6 kg |
| Flavours and colouring | To taste |

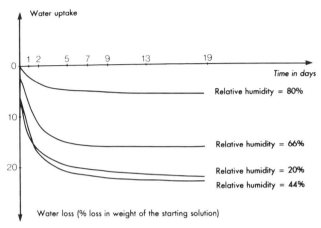

FIG. 7.   Loss of water from a solution of Lycasin 80/55 originally at 70 % solids at different relative humidities.

Boiling is carried out at 135 °C to a final weight of 100 kg, and the mixture is beaten. Moisture-proof packaging must be provided.

### 4.4.4.  Jellies and Pastilles

Jellified confectionery can be based on Lycasin 80/55, the products being coated with mannitol which is not hygroscopic. For items containing gelatin it is important to use a 220 Bloom limed bone gelatin and in particular to maintain at neutral pH to obtain good gelling. The 220 Bloom gelatin (9 kg) is soaked in lukewarm water (20 kg) for 30 min. The Lycasin 80/55 (120 kg) is boiled at 113 °C to obtain an 85 % solids syrup and then cooled to 65–70 °C, at which temperature the soaked gelatin, flavours and colouring (to taste) are added. After settling for 30 min and removal of froth, the mixture is deposited on a very dry cold starch support and left for 24 h.

A typical recipe for pectin-based articles is as follows:

| | |
|---|---|
| Pectin (Unipectin yellow ribbon type) | 1·7 |
| Water | 30 |
| Lycasin 80/55 | 95 |
| Tartaric acid | to taste |
| Flavours and colouring | to taste |

The pectin is dispersed in water and then, after heating to boiling, the Lycasin 80/55 is added and the mixture is heated to approximately 108 °C.

After adding the other components, it is deposited in dry starch and left for a maximum of 4 h.

For gum-arabic-based articles, Lycasin 80/55 (34 parts) is first cooked at 135 °C. After cooling to 65 °C, 50 parts of 50 % gum arabic solution are added. The mixture is deposited in dry warm starch and stoved for 48 h at 50–60 °C.

### 4.4.5. Coating

After concentration to 85 % solids, Lycasin 80/55 can be used for soft coatings in conjunction with sorbitol powder.

## 5. CONCLUSIONS

This chapter has deliberately been limited to applications in the food industry and the technological characteristics relating to them. Hydrogenated derivatives have, in addition, very many applications in a wide range of fields, such as pharmacy (e.g. as excipients and in products for injection), chemistry (e.g. the synthesis of vitamin C and of surfactants), animal feeds, paper manufacture, glues and textiles.

For food applications, consideration naturally has to be given in each instance to the regulations covering the use and labelling of the particular range of products, which can vary from one country to another.

## REFERENCES

1. JEROME, N. W. (1977). In: Carbohydrates and Health, L. Hood, E. Wardrip and G. Bollenback (Eds.), Avi Publishing Co. Inc., Westport, Conn., 1–2.
2. BERNIER, J. J. and PAUPE, J. (1963). Les Glucides, Masson, Paris, 9.
3. ROCKSTRÖM, E. (1980). In: Carbohydrate Sweeteners in Foods and Nutrition, P. Koivistoinen and L. Hyvönen (Eds.), Academic Press, London, 225–32.
4. CROWLEY, M., HARNER, V., BENNET, A. and JAY, P. (1956). J. Amer. Dent. Assoc., 52, 148.
5. LEROY, P. (1978). In: Health and Sugar Substitutes, B. Guggenheim (Ed.), Proc. ERGOB Conf. Geneva, S. Karger, Basel, 114.
6. PROUST, L. (1806). Ann. Chim. Phys., 57, 1.
7. SORENSEN, N. A. and KRISTENSEN, K. (1950). US Patent 2 516 350.
8. BIRKENSHAW, J. H. (1931). Trans. R. Soc., London, B, 220, 153.
9. YAMASAKI, I. and SHIMOMURA, M. (1977). Biochem. Z, 291, 340.
10. SMILEY, K., CADMUS, M. and ROGORIN, S. P. (1969). US Patent 3 427 224.
11. ONISHI, H. and SUZUKI, T. (1971). US Patent 3 622 456.

12. BEHRENS, V., SATTLER, K. and WUNSCHE, L. (1967). German Patent No. 1 296 613.
13. WÜNSCHE, L., SATTLER, K. and BEERENS, V. (1966). *Zeit. All. Mikrobiol.*, **6**, 323.
14. BRANDNER, J. D. and WRIGHT, L. W. (1962). US Patent 3 329 729.
15. PIJNENBURG, H. C. M., KUSTER, B. F. M. and VAN DER BAAN, H. S. (1978). *Stärke*, **30**, 199–205.
16. ICI (1975). British Patent No. 1 506 534.
17. DERMER, J. (1946). *Proc. Okla. Acad. Sci.*, **27**, 9–20.
18. ALTAN, A. (1967). US Patent 3 632 656.
19. HALES, A. R. (1967). US Patent 3 484 492.
20. BOUSSINGAULT, J. (1872). *Compt. Rend.*, **74**, 939.
21. BRIMACOMBE, J. S. and WEBBER, J. M. (1974). *Carbohydrates—Chemistry and Biochemistry*, Academic Press, New York, Vol. 1A, 487.
22. TAYLOR, R. L. (1937). *Chem. Met. Eng.*, **44**, 588.
23. SANDERS, M. T. and HALES, R. A. (1949). *J. Electrochem. Soc.*, **96**, 241.
24. *Official Journal of the European Communities* (1978). L 223/14–15, 14 August, (Council Directive 78/663/EEC).
25. ROSE, R. S. and GOEPP, R. M. (1939). *The Determination of Some Physical Constants of Sorbitol—the Polymorphism of Sorbitol*. Atlas Powder Co., Amer. Chem. Soc. Meeting, Baltimore, MA.
26. MANGIN, H. and HUCHETTE, M. (1972). French Patent No. 72 36437.
27. FURIA, T. E. (Ed.) (1972). *Handbook of Food Additives*, 2nd Edn., CRC Press, Cleveland, Ohio.
28. FDA Report No. PB 221 210 (1972), 5.
29. VERWAERDE, F., LELEU, J. B. and HUCHETTE, M. (1979). Patent application GB 2 038 832.
30. IMFIELD, T. and MUHLEMANN, H. R. (1978). *Caries Res.*, **12**, 256.

Chapter 2

# LACTOSE HYDROLYSATE SYRUPS: PHYSIOLOGICAL AND METABOLIC EFFECTS

CELIA A. WILLIAMS

*Department of Biochemistry, University of Surrey,
Guildford, Surrey, UK*

*SUMMARY*

*Lactose hydrolysate syrups containing the monosaccharides galactose and glucose as well as some lactose are produced commercially from lactose. Some effects produced from ingestion of these carbohydrates are described. Chronically high serum levels of galactose are undesirable, as demonstrated in the congenital disorder of clinical galactosaemia. Serum galactose levels are reduced when glucose is taken with galactose. Before absorption, lactose has to be hydrolysed in the small intestine. A large proportion of adults have a limited capacity to hydrolyse lactose and are therefore lactose-intolerant; however, they can metabolise a mixture of galactose and glucose as efficiently as lactose-tolerant individuals. The toxicity of galactose has been repeatedly demonstrated in rats, and it is concluded that the rat is a poor model for studying galactose metabolism in man.*

## 1. INTRODUCTION

Lactose is a disaccharide found only in milk. It has low water solubility and sweetness which limit its suitability for use in many food products. Hydrolysis of lactose makes it more soluble and sweeter, and produces the monosaccharides galactose and glucose, which are epimeric aldo-hexoses

27

differing only in the configuration of the hydroxyl group on carbon atom 4 (Fig. 1). Hydrolysis of lactose takes place in the small intestine in the presence of the enzyme lactase. Lactose can be hydrolysed commercially, either enzymically or by acid, to produce lactose hydrolysate syrups which contain lactose, galactose and glucose. The solubility of lactose at room temperature is low, 18 % w/v, compared with that of glucose, 50 % w/v, and galactose, 32 % w/v. The optimum solubility of a lactose hydrolysate is

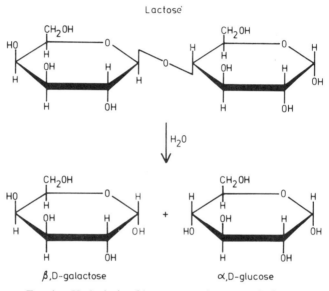

FIG. 1.   Hydrolysis of lactose to galactose and glucose.

achieved with 75 % hydrolysis of the lactose. The sweetness of lactose lies between 15 and 30 relative to sucrose (reference value 100). Under similar conditions the sweetness of galactose is 40–50 and glucose 60–75, both monosaccharides being significantly sweeter than lactose. In lactose hydrolysate syrups the three sugars act synergistically and produce a sweeter product than would be expected from the sum of their proportionate sweetness intensities. The properties and applications of lactose hydrolysate syrups have been reviewed recently.[1] Possible uses include the replacement of sucrose in dairy products, such as ice-cream and yoghurt, and also in bakery products, confectionery and beer.

There has been very little work published on the physiological and metabolic effects of lactose hydrolysate syrup preparations in man, but some of the effects of galactose and glucose given as individual monosaccharides, or as a mixture, and the effects of lactose have been described and will be reviewed in this article.

## 2. EFFECT ON GASTRIC EMPTYING

Ingestion of osmotically active solutions can slow gastric emptying,[2] and this effect is believed to be due to the excitation of osmoreceptors in the duodenum. In man, galactose has been found to be less effective at slowing gastric emptying than glucose, although their molecular weights, and hence osmotic activities, are the same.[3] A mixture of galactose and glucose has the same effect as galactose alone on gastric emptying,[3] and lactose has been shown to have the same effect as a mixture of galactose and glucose[4] in lactose-tolerant subjects. Lactose hydrolysate syrup in the stomach may then affect gastric emptying as galactose would, i.e. to a lesser extent than glucose.

## 3. INTESTINAL HYDROLYSIS OF LACTOSE

Disaccharides are not absorbed intact by man but prior to absorption are hydrolysed by specific disaccharidases to their constituent mono-saccharides which can be absorbed directly. Lactose is hydrolysed to galactose and glucose by the enzyme lactase in the brush border of the mucosal epithelium cells. The activity of lactase is highest in the distal duodenum and jejunum.[5] Lactase is one of three $\beta$-galactosidases found in the small intestine. Acid $\beta$-galactosidase, in lysosomes, is capable of hydrolysing lactose, while hetero-$\beta$-galactosidase, in cytoplasm, cannot hydrolyse lactose.

It is believed that the rate of hydrolysis of lactose does not limit the total absorption of galactose and glucose in lactose-tolerant individuals.[6,7] However, the absorption of galactose and glucose from lactose has been found to be slower than the absorption of galactose and glucose from a mixture of the monosaccharides.[8] This is unlike the absorption rates of the monosaccharides derived from sucrose and maltose, which are the same[6,9,10] or faster than those of the monosaccharide.[11] Most of these studies involved jejunal perfusion techniques, the limitations of which are discussed later.

There are three recognised types of lactase deficiency in which the subject with reduced lactase activity is said to be lactose-intolerant. Undigested lactose exerts an osmotic effect, drawing water into the intestinal lumen, and travels to the large intestine where the bacterial flora of the colon ferment the lactose to various acids and carbon dioxide. Depending on the amount of lactose consumed, the subject with a reduced lactase activity can experience anything from a mildly bloated feeling to severe diarrhoea.

Congenital lactase deficiency is a rare disorder, first described by Holzel et al.,[12] where lactase is absent from birth due to mutation of a structural gene. If it is not diagnosed promptly it can be fatal. Secondary lactase deficiency occurs as a consequence of atrophy of the mucosa of the small intestine. Usually the activity of other disaccharidases is similarly affected and on recovery the enzymes reappear.

In primary lactase deficiency, first described by Dahlqvist et al.,[13] lactase is not deficient at birth but in adulthood the subject shows reduced lactase activity and impaired lactose absorption. In 1965 Cuatrecasas et al.[14] gave lactose to both Black and White Americans, and found a significantly higher proportion of the Black Americans to be lactose non-absorbers with a reduced lactase activity compared to White Americans. A racial difference in the ability to tolerate lactose was confirmed by Cook and Kajubi[15] and by Bayless and Rosenweig.[16] Subsequent research has shown that lactose intolerance predominates in many races including Black Africans, North American Indians, Eskimos, Greek Cypriots and the people of the Pacific Isles.[17–19]

Lafayette and Mendel[20] first noted the decline of lactase activity in animals subsequent to weaning. It is now established that in all mammals and in 60–90 % of non-Caucasians lactase activity is highest in the neonate and falls to low levels after weaning. The notable exceptions are the 85–95 % of Caucasians who maintain a high level of lactase activity in adulthood,[17] and, amongst the mammals, the Pinnipedia, a group including sealions, seals and walruses, which have no lactase activity even in infancy.[21]

Currently two explanations are put forward for this phenomenon. The first is based on the fact that most individuals who tolerate lactose have a high history of drinking milk after weaning. Since enzyme adaption to an environmental stimulus is well documented, this tolerance would not be genetically determined. The second explanation is based on the Darwinian concept of evolution that, if lactase persistance proved to be of biological advantage, it would be transferred genetically within populations.[18,19,22–5] According to this theory a reduced level of lactase activity in adults is the natural physiological state. The environmental theory has been the subject of much criticism because the majority of experiments in man and animals have failed to induce lactase.[14,26-30] Similarly, studies in lactose-intolerant populations have failed to correlate lactose intolerance with duration of breast feeding, continued milk consumption, previous malnutrition or present dietary status.[31-5] Perhaps the strongest piece of evidence in support of the genetic-alteration theory, and disproving the environmental-adaptation theory, is that the incidence of lactose-intolerance among African Negroes is 72%, and among their American counterparts, with a different dietary tradition extending over 300 years, the incidence of intolerance is still 70%.[18]

The lactose-intolerant subject need not necessarily abstain from milk consumption[14,31,33] and can take up to 950 ml per day in small quantities.[36] Food accompanying milk retains its nutritional value in the lactase-deficient person.[37] Although lactose can be ingested by lactase-deficient subjects, the amount of galactose and glucose absorbed from it would be very small.

## 4. ABSORPTION OF GALACTOSE AND GLUCOSE

Monosaccharides are absorbed from the small intestine at different rates. In 1925, Cori,[38] using rats, obtained absorption coefficients for a range of monosaccharides, and rated glucose as 100, galactose as 110, fructose as 43 and the others below 20. In vivo studies in man produce similar coefficients: glucose 100, galactose 122 and fructose 67.[39] From these early findings it was apparent that galactose and glucose are absorbed more efficiently than the other monosaccharides. There have been numerous studies concerned with the mechanism of monosaccharide absorption—these were reviewed by Crane[40] who suggested a model which is supported by current experimental evidence.[41] In the brush border of the intestinal mucosa there is a sodium-dependent carrier which is shared by galactose and glucose. It is possible that there may be more than one such carrier;[42,43] interaction between the carrier and the localised sodium pump allows the monosaccharide to be actively concentrated within the cell. This process is known as active transfer, and galactose and glucose are said to be actively absorbed—that is, their transfer across the intestinal mucosal epithelium can occur against a concentration gradient and is an energy-dependent process whereas other monosaccharides are absorbed by passive diffusion. The model of galactose and glucose sharing a carrier in the intestine is supported by clinical evidence; there is a congenital condition of galactose–glucose malabsorption in which the subject cannot efficiently absorb either galactose or glucose whereas fructose absorption is unaffected.[44-6]

Recent studies in man have used a jejunal perfusion technique whereby a segment of jejunum can be perfused through an accurately positioned tube and samples of perfusate removed perorally. Such studies have confirmed that the maximum rate of galactose absorption is higher than that of glucose absorption.[47,48] However, a lower concentration of glucose is required in the intestinal lumen to produce a maximum rate of transfer,[47,49] suggesting that the carrier may have an increased affinity for glucose.

Since galactose and glucose share the same intestinal transport system, mutual inhibition of the absorption of both monosaccharides might be expected on administration of an equimolar mixture of galactose and glucose as found in a lactose hydrolysate syrup. Values calculated from kinetic investigations of galactose and glucose absorption in man predict the inhibition of galactose absorption by glucose to be 30 % and the inhibition of glucose absorption by galactose to be 21 %.[50] Results from a study in man using the jejunal perfusion technique showed galactose absorption to be inhibited by 29 % in the presence of glucose, as predicted, yet glucose absorption was found to be unaffected by galactose.[47] Similarly glucose absorption has been shown to be more rapid than galactose absorption during either galactose and glucose or lactose perfusion.[8] These results were all obtained with perfusion techniques which involved only small portions of the intestine. Studies in which galactose and glucose mixtures were infused into the whole intestine of calves have shown glucose absorption to occur rapidly in the proximal intestine and galactose absorption to occur in the distal intestine.[51] The values obtained from perfusion of specific intestinal segments may then be of only limited value and may not give a quantitative picture of either absorption or hydrolysis.

On physiological grounds, it is unlikely that ingestion of a galactose–glucose mixture containing a small percentage of lactose (i.e. a lactose hydrolysate syrup) would cause intestinal discomfort due to malabsorption in lactose-tolerant humans because both monosaccharides are absorbed efficiently.

## 5.  METABOLISM OF GALACTOSE AND GLUCOSE

The major metabolic function of carbohydrate is to provide the energy essential for all biological processes including maintenance of a normal physiological and metabolic state, work and muscle activity, growth, reproduction and thermal regulation. Dietary carbohydrate can be considered to be a fuel source, and as such it is utilised at the cellular level in the form of glucose. Fructose and galactose, the only other readily absorbed monosaccharides, are converted to glucose in the liver. Under normal conditions glucose is the essential energy source for the brain erythrocytes, and it has been estimated that 180 g of glucose is needed per day by a normal adult for this purpose.[52] It is only during starvation that the brain adapts to metabolise fatty acids. Although carbohydrate can be metabolised to fat, the reverse is not possible; if carbohydrate is not available in the diet and body stores are depleted, carbohydrate can be synthesised at the expense of protein. A certain amount of carbohydrate is essential to prevent ketosis, since carbohydrate can be metabolised to oxaloacetate which is used in the further metabolism of ketone bodies. Details of the biochemistry and regulation of glucose and galactose metabolism are well known. A brief outline is given in Fig. 2.

Glucose is a readily transportable substrate, and prior to its entry into either synthetic or degradative pathways it must be phosphorylated to glucose-6-phosphate (glc-6-p), a reaction requiring 1 mole of adenosine triphosphate (ATP) per mole of glucose. The reaction can be catalysed by either hexokinase, which is distributed widely through the tissues, or by glucokinase, which is found specifically in the liver and has different kinetic properties from hexokinase which allow it to efficiently remove, from the blood, glucose entering the circulation from the intestine. The reverse reaction, i.e. the dephosphorylation of glucose, is catalysed by glucose-6-phosphatase, and is essential prior to its release from the cell. Glc-6-p can then either enter the oxidative processes of glycolysis or the pentose phosphate pathway to provide chemical energy in the form of ATP, or it can be used to synthesise glycogen.

Glycolysis involves the oxidation of glc-6-p to lactate under anaerobic conditions or to pyruvate under aerobic conditions, with a net gain of ATP. Lactate can be oxidised to pyruvate under aerobic conditions, and pyruvate can be metabolised to acetyl-CoA which can enter the Krebs or citric acid cycle, the final common pathway of oxidation for carbohydrate, fat and protein. Acetyl-CoA is completely oxidised to carbon dioxide and water with the simultaneous generation of ATP.

In the process of both the glycolytic pathway and the citric acid cycle, the intermediate steps produce precursors for other synthetic reactions. During glycolysis dihydroxyacetone phosphate is formed which can be metabolised to α-glycerophosphate—required for triglyceride synthesis.

Glc-6-p can also enter the pentose phosphate pathway, an alternative route for the oxidation of glucose. This pathway supplies a reduced coenzyme which is necessary for synthetic reactions such as fatty acid synthesis.

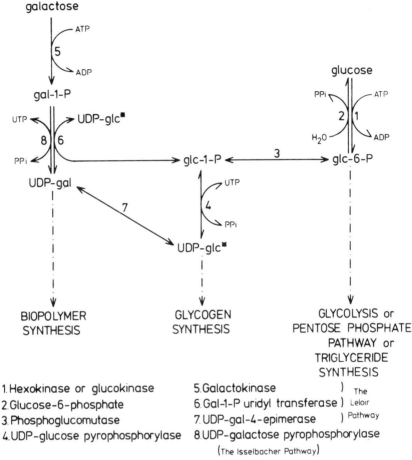

FIG. 2.    The interrelationship of galactose and glucose metabolism.

1. Hexokinase or glucokinase
2. Glucose-6-phosphate
3. Phosphoglucomutase
4. UDP-glucose pyrophosphorylase

5. Galactokinase
6. Gal-1-P uridyl transferase
7. UDP-gal-4-epimerase
8. UDP-galactose pyrophosphorylase

) The
) Leloir
) Pathway

(The Isselbacher Pathway)

A limited amount of carbohydrate can be stored as glycogen in the liver and muscle. Glycogen is synthesised from glucose via glc-6-p, which is converted to glucose-1-phosphate (glc-1-p) by phosphoglucomutase; glucose cannot be metabolised directly to glc-1-p. In the presence of the enzyme UDP-glucose pyrophosphorylase, glc-1-p is metabolised to uridine diphosphate glucose (UDP-glc) utilising uridine triphosphate (UTP). The UDP-glc then transfers the glucose to a glycogen chain with the liberation of UDP. Glucogen synthesis is finely controlled by hormones. The glycogen stored in the liver is used to maintain blood glucose levels.

However, the glycogen of muscle cannot be released as glucose and is used as an energy source when required by the cells in which it is synthesised. UDP-glc is part of a group of metabolites called the nucleoside diphosphate sugars which are involved in the synthesis of polysaccharides, e.g. glycogen, and of disaccharides although they can also be used in the interconversion of monosaccharides. It is through their UDP derivatives that glucose and galactose metabolism become interconnected. UDP-glc can be involved in the metabolism of galactose to uridine diphosphate galactose (UDP-gal). UDP-gal and UDP-glc are interconverted by the enzyme UDP-galactose-4-epimerase, a reaction studied in detail by Leloir who gave his name to this major route of galactose metabolism.[53]

The Leloir pathway, the main route of galactose metabolism, involves four steps and results in the overall conversion of galactose to glc-1-p.[54-6] The primary site of the Leloir pathway is the liver but it is also found in the kidney and erythrocytes.[55] Galactose is phosphorylated by galactokinase to galactose-1-phosphate (gal-1-p) at the expense of ATP. Gal-1-p and UDP-glc react in the presence of galactose-1-phosphate uridyl transferase to produce UDP-gal and glc-1-p. (An alternative route for the synthesis of UDP-gal is mentioned later.) The glc-1-p is then converted to glc-6-p and metabolised as described for glucose. UDP-galactose is either used for synthesis or converted to UDP-glucose by UDP-galactose-4-epimerase.

Galactose is an essential component of gangliosides, cerebrosides found in the brain and nervous tissue, and of some phospholipids.[57,58] Galactose is incorporated into these biopolymers in reactions involving UDP-gal. During lactation UDP-gal is involved in lactose synthesis. The UDP-glc produced from UDP-gal can be used in Step 2 of the Leloir pathway in glycogen or glc-1-p synthesis. The interconversion of UDP-gal and UDP-glc is of anabolic importance because normally the quantity of ingested galactose does not meet synthetic requirements. All the UDP-gal required during lactation is derived from blood glucose.

The Isselbacher pathway is an alternative route for Step 2 of the Leloir pathway. Gal-1-p and UTP react in the presence of UDP-galactose pyrophosphorylase to produce UDP-gal and inorganic phosphate.[59]

## 6. METABOLIC EFFECTS OF GALACTOSE, GLUCOSE AND LACTOSE

The physiology of both galactose and glucose metabolism as individual monosaccharides is very different. Glucose is the physiological mono-saccharide for man and is metabolised by all the tissues of the body. Blood glucose levels are maintained within well-defined limits; fasting levels are normally between 70 and 100 mg 100 ml$^{-1}$ and rarely rise above 140 mg 100 ml$^{-1}$ even after a carbohydrate-rich meal. Deviations outside these limits have immediate widespread metabolic effects. Insulin and glucagon, hormones produced in and secreted from the pancreas, are involved in blood glucose regulation. Insulin, secreted in response to a rise in blood glucose concentration, increases the rate of glucose metabolism and glycogen synthesis and also the permeability of muscle cells and adipocytes to glucose, resulting in a fall in blood glucose levels. Glucagon, which acts antagonistically to insulin, is secreted in response to a fall in blood glucose levels and stimulates the breakdown of glycogen and glucose synthesis, and thus effects an increase in blood glucose concentration. It is because of this regulation that the serum glucose response to a glucose meal does not vary significantly over a wide range of ingested glucose loads.[60-3] In normal humans glucose is rarely found in the urine subsequent to its ingestion because reabsorption of glucose in the nephron of the kidney subsequent to its loss by diffusion is very efficient.

Galactose is metabolised to a significant extent by the liver alone. Because of this, a galactose-tolerance test has been used diagnostically to assess liver function.[64-8] The erythrocytes and the kidney are capable of metabolising galactose in man[55] but the extent of their contribution to the overall metabolism of galactose is unknown. Fasting serum galactose concentration is normally below 4·3 mg 100 ml$^{-1}$,[69] but, unlike the serum glucose response to glucose, the serum galactose response to galactose is dependent upon the amount ingested. The maximum concentration of galactose in the blood and the duration of the elevated serum galactose level increase with the size of the galactose load.[66] As blood galactose levels are uncontrolled, galactose is often found in the urine after its ingestion— leading to high blood levels—and can be found in the urine at plasma levels as low as 9 mg 100 ml$^{-1}$.[70] Reports on the effect of insulin on galactose metabolism in man and in other animals are conflicting;[71-5] however, since the serum galactose responses to peroral galactose in diabetic and non-diabetic humans are the same,[71] insulin is not essential for a normal response.

### 6.1. Serum Galactose

In man, it has been shown that ingestion of glucose with a galactose meal reduces the subsequent serum galactose response.[66,76] This effect is related to the quantity of glucose accompanying the galactose meal.[76] Figure 3 shows the mean serum galactose response of human volunteers given 0·5 g galactose per kilogram body weight in the form of either galactose, galactose plus glucose, or lactose. The presence of glucose in the test meal, either as such or as part of the dissaccharide lactose, considerably reduces the serum galactose response. Addition of glucose to a galactose meal also reduces the loss of galactose in the urine.[78-80] Since lactose hydrolysate

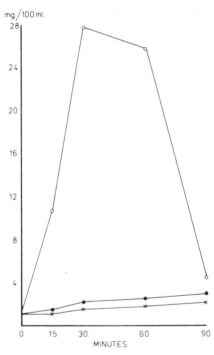

FIG. 3.  The mean serum galactose response of 12 fasted human volunteers to ingestion of either 0·5 g galactose per kilogram body weight alone or with 0·5 g glucose per kilogram body weight, or 1·0 g lactose per kilogram body weight.[77]
○, Galactose; ●, glucose + galactose; ×, lactose.

syrup contains equal amounts of galactose and glucose in either mono-saccharide or disaccharide form, it can be assumed that the serum galactose response to lactose-hydrolysate-containing foods will be low due to the presence of glucose.

The mechanism by which glucose reduces the serum galactose response to peroral galactose is unknown. It was initially believed to be due to the two monosaccharides competing for absorption from the intestinal lumen;[81,82] however, if galactose is given perorally and glucose intra-venously, thus bypassing the intestinal lumen, the serum galactose response is still reduced.[66,76,83]

## 6.2. Serum Glucose

Ingestion of galactose by man produces an increase in the serum glucose concentration (Fig. 4)[66,76,84,85] which is believed to be due to an increase in the hepatic output of glucose. Studies in normal subjects using an intravenous infusion of galactose[1-$C^{14}$] show the amount of $^{14}CO_2$ produced to be only slightly less than that produced after glucose[1-$C^{14}$] in the fasting state, and from estimation of the glucose pools and measurement of the specific activity of glucose it has been shown that 30 min after injection of galactose 45 % is in the body glucose pools.[86] This

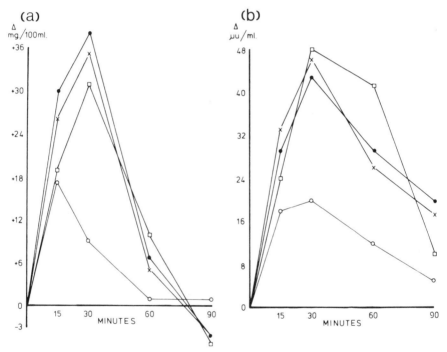

FIG. 4. (a) The mean serum glucose, and (b) mean serum insulin response of 12 fasted human volunteers to ingestion of either 0·5 g galactose per kilogram body weight alone or with 0·5 g glucose per kilogram body weight, 0·5 g glucose per kilogram body weight alone, or 1·0 g lactose per kilogram body weight.[77] □, Glucose; ○, galactose; ●, glucose + galactose; ×, lactose.

demonstrates the capacity of the liver for the conversion of galactose to glucose. The serum glucose response to a galactose plus glucose mixture and to lactose is the same as that produced by ingestion of an equivalent glucose load.

### 6.3. Serum Insulin

Peroral galactose produces a slight increase in serum insulin (Fig. 4).[83,85,87,88] Galactose is believed to be non-insulinogenic in man[89-91] and other animals,[92-5] the increase in the serum glucose level following peroral galactose intake being responsible for the rise in serum insulin levels. Meals of galactose plus glucose and of lactose produce similar serum insulin levels which only differ from the serum insulin levels after ingestion of an equivalent amount of glucose alone at 60 and 90 min.

### 6.4. Respiratory Quotient (RQ)

The RQ is defined as the ratio of carbon dioxide expired to oxygen consumed over a given period of time. This ratio alters according to the substrate or fuel type being oxidised by the body; if carbohydrate is being utilised the RQ is 1, while if fat is being utilised the RQ is 0·7. Various carbohydrates can produce different effects on the RQ in man. Ingestion of galactose causes a more rapid rise in RQ than does ingestion of glucose,[82,96,97] suggesting that galactose is probably oxidised or converted to fat more readily than glucose.[82] The rise in RQ produced by glucose is sustained longer than that produced by galactose.[97] A mixture of galactose and glucose also produces a rapid rise in RQ, as does galactose alone, whereas the duration of the increase in RQ is prolonged as after glucose. Lactose produces a rapid increase in RQ and follows the same pattern of RQ response as is seen after galactose plus glucose, although the RQ is lower after lactose than after the monosaccharides.[97] These differences in RQ indicate that the monosaccharides galactose and glucose have different metabolic effects whereas galactose plus glucose and lactose have very similar effects.

### 6.5. Calcium Metabolism

Lactose has long been associated with calcium metabolism in man and other animals.[98] Lactose has been reported to increase the absorption of calcium from the small intestine but the mechanism by which this effect occurs is not understood.[99,100] A 1980 WHO report[101] found the results from human and animal studies on the effect of lactose on calcium absorption to be inconsistent. Similarly there are conflicting reports on the effect of galactose on calcium absorption.[102-4] It has been found in man that the beneficial effect of lactose on intestinal calcium absorption is dependent on the ability of the individual to hydrolyse lactose.[105] The absorption and retention of many other minerals, including lead, magnesium and iron, have been shown to be improved by dietary lactose, and it has been suggested that lactose has a non-specific effect on the intestinal absorption of minerals.[106] Although the relationship between lactose or galactose and the absorption of calcium is not yet resolved, it should be noted that milk, the sole source of lactose, is also a rich calcium source for young mammals.

### 6.6. Other Effects

Serum lactate and pyruvate levels increase in response to peroral galactose in fasting humans when galactose alone[91, 107, 108] or galactose with glucose or lactose is fed.[108] Normally there is no increase in serum lactate concentration after ingestion of galactose;[91] peroral glucose, however, causes a fall in serum lactate levels and does not significantly affect serum pyruvate levels.[63]

Serum uric acid concentrations increase after galactose ingestion[108] whereas glucose has no effect.[63] The effects of peroral galactose on serum lactate, pyruvate and uric acid levels are similar to those of fructose.

Studies in rats and dogs have indicated that glycogen is formed less readily from galactose than from glucose or fructose,[82,109,110] but these results cannot be extrapolated to man because rats and dogs have a lower tolerance to galactose.[111,112]

### 6.7. Factors Affecting the Metabolic Response to Galactose

Since an elevated serum galactose concentration is undesirable (see Section 8), factors affecting galactose metabolism must be considered. The effect of glucose on galactose metabolism has already been described. The effect of alcohol on galactose metabolism is of interest. In normal humans, ethanol reduces tolerance to galactose,[113,114] and increasing doses of alcohol increase the duration of galactosaemia.[66,115] Studies using $^{14}$C-labelled galactose have shown that ethanol inhibits the metabolism of galactose to carbon dioxide in humans.[86] Ethanol oxidation increases the NADH: $NAD^+$ ratio which inhibits UDP-galactose-4-epimerase, thus reducing the hepatic uptake of galactose.[116]

Carbohydrates can bring about adaptive changes in the activities of enzymes concerned with carbohydrate metabolism—an absence of a specific carbohydrate from the diet resulting in the decline of the activity of enzymes concerned with its metabolism.[117] However, the mean serum galactose, glucose and insulin responses of six lactose-intolerant subjects to a galactose and glucose mixture were the same as those found normally.[118] Lactose intolerance does not appear to diminish the ability to metabolise a galactose load ingested with glucose. This is of significance if food products containing lactose hydrolysate syrup are to become available to lactose-intolerant populations.

## 7. CHRONIC CONSUMPTION OF GALACTOSE AND GLUCOSE

The effects described so far have been acute effects, i.e. produced in the first hour or two after ingestion of a test meal. There has been no specific study of the metabolic effects produced by chronic consumption of lactose hydrolysate syrup in man. However, two diets containing approximately 80 % by weight lactose hydrolysate syrup, one produced by acid hydrolysis and the other by enzyme hydrolysis, have been fed to baboons over a 10-week dietary period.[80] The most notable effect of both diets was on the fasting serum triglyceride concentrations of the female baboons. During the later stages of both the lactose hydrolysate syrup dietary periods the mean fasting serum triglyceride level was significantly elevated in females— an effect not seen in males.[80] Another study has reported an increase in fasting serum triglyceride concentrations of 2 out of 3 female and 1 out of 3 male baboons with a diet containing 40 % by weight lactose.[119] These results are pertinent because the baboon is considered to be a good model of lipid metabolism for extrapolation to man,[120-2] and also because the baboon, like man, can develop atherosclerosis in its natural environment.[123-5] An increased intake of carbohydrate has been found to produce a transient increase in the fasting serum triglyceride concentrations of man.[126-8] The lipogenic effect of carbohydrate is of interest because a chronically elevated fasting serum triglyceride level has been related to coronary artery disease.[129-31] The effect of carbohydrate on lipid metabolism varies according to the type of carbohydrate,[132,133] but the lipogenic effect of carbohydrate is usually seen in the male of the species.[134-6] The effect of lactose hydrolysate diets on lipid metabolism in female baboons is thus of particular interest and warrants further study.

## 8.  GALACTOSE TOXICITY

The focus of much research in the study of galactose metabolism has been the possible hazardous nature of this monosaccharide as demonstrated in clinical galactosaemia and galactose-induced cataract.

### 8.1.  Galactosaemia

There are two clinically recognised forms of galactosaemia which occur as inborn errors of metabolism. In the absence of an enzyme concerned with the normal metabolism of galactose, ingestion of galactose or lactose results in a chronically increased serum galactose level and subsequent excretion and/or accumulation of galactose and galactose derivatives. Classical galactosaemia is due to galactose-1-phosphate uridyl transferase deficiency,[137] a consequence of which is that absorbed galactose can be phosphorylated to galactose-1-phosphate but not metabolised further, resulting in an accumulation of galactose-1-phosphate in the tissues.[138,139] Symptoms of classical galactosaemia include liver malfunction, cataracts, failure to thrive, and mental retardation. An early elimination of galactose from the diet is essential to normal development.

In galactokinase-deficient galactosaemia, absorbed galactose is not phosphorylated. The condition is characterised by cataracts which become apparent in the first and second decades of life but the subject is otherwise normal.[140] Both types of galactosaemia, their implications, nature of inheritance and the newly discovered variants are fully described in the literature.[55,141-3]

The contrast between the clinical manifestations of the two forms of galactosaemia demonstrates the toxicity of an excess of galactose-1-phosphate in man. Cataract, common to both disorders, is caused by unphosphorylated galactose. At the cellular level galactose-1-phosphate can interfere with glucose metabolism, the pentose phosphate pathway and glycogen formation by inhibiting a number of enzymes.[141,143] Both types of galactosaemia can be controlled by a galactose-free diet.

### 8.2.  Cataract

Much of the research concerning galactose cataract has been carried out in rats. It was first demonstrated that a diet high in lactose or galactose could produce cataract in rats[144] and subsequent studies have shown cataracts to form in rats fed diets containing 30% or more of their total energy as galactose.[145-8] A diet containing 22% of its total energy as galactose, based on yoghurt, caused cataract to develop in rats which otherwise grew and reproduced normally.[149] In 1972, a study to determine the dose-dependence of cataract in rat found that a diet containing as little as 15% of its total energy in the form of galactose was capable of producing cortical cataract in rats.[150] It has also been found that younger rats are more susceptible to cataract[149] and that cataract can be induced in the foetuses of rats by feeding the dams high-galactose diets.[146,151] Direct extrapolation of these quantitative results to man is not possible since the rat has been found to have a lower tolerance to galactose than man[111,112] and the galactose utilisation rate of the rat liver is only half that found in man.[152] Also, glucose does not reduce the serum galactose response to ingested galactose in rats, as it does in man.[153]

Riboflavin deficiency in rats has been found to promote the development of galactose-induced cataracts,[154] although in man tolerance of galactose is unaffected by riboflavin status.[155]

The pathogenesis of galactose cataract has also been studied extensively in rats.[145,147,156-9] Formation of galactose cataract is believed to be due to an excess of galactose entering the lens from the blood and being metabolised to galactitol (dulcitol) in the presence of aldose reductase.[160] Once converted to galactitol, the monosaccharide alcohol cannot leave the lens and so accumulates, exerting osmotic pressure within the lens. Thus water can then enter, disrupt the structure of the lens and lead to an imbalance in electrolytes, amino acids and proteins.[161-3] The development of galactose cataract in man is believed to occur in the same way.

Although the rat is not a good model for quantitative studies of galactose cataract, the studies indicate that a chronically elevated serum galactose level in man would be undesirable.

Galactose is not the only monosaccharide that can cause cataract; glucose, xylose and galactose all have been called the 'cataractogenic monosaccharides' in a recent review.[164] Galactose is probably of the greatest potential danger to normal subjects, as not only is it actively absorbed but, unlike glucose, the magnitude and duration of the ensuing galactosaemia is dose related. Glucose can only cause cataract in diabetic models. In normal humans blood glucose levels are maintained within narrow limits primarily by the action of insulin; xylose is not actively absorbed and if an excess is ingested it remains unabsorbed causing an osmotic diarrhoea. In lactose hydrolysate syrups galactose is present with an equal amount of glucose and, as already mentioned, ingestion of a mixture of galactose and glucose results in considerably lower blood galactose levels in man. The chronically elevated blood galactose levels involved in the pathogenesis of galactose cataract are not developed.

**8.3. Other Aspects of Galactose Toxicity**

There is evidence that foetuses from female rats receiving high-galactose diets during pregnancy show growth retardation, cataracts and lesions of the liver and kidney.[146,151,165,166] Chicks fed galactose for 2–3 days develop tremor and violent convulsions, and their brains appear to be deprived of glucose.[141] The rat is a poor model for galactose metabolism in man and the chick must also be considered an inadequate model as it would never ingest galactose naturally.

## 9. CONCLUSIONS

The acute metabolic effects described in this chapter are produced on ingestion of galactose with or without glucose and lactose in solution, and not those produced on ingestion of lactose hydrolysate syrup. The quantities of the monosaccharides and disaccharides given experimentally are larger than would be encountered in diets including lactose-hydrolysate-syrup-containing foods. The medium in which the syrup is presented would also affect the immediate physiological effect—for example, its pH and osmolarity would affect gastric emptying, and its viscosity may alter the availability of the monosaccharides to the absorptive surface and the availability of the disaccharide to disaccharidase.

Although lactose is absorbed as galactose and glucose, it is possible that it has some metabolic effects different from those produced on ingestion of a mixture of glucose and galactose. This is evident in the case of calcium metabolism and in lactose's effects on RQ; prior to absorption lactose may possibly produce different effects on the secretion of a gastrointestinal hormone(s). Similarly, although galactose is efficiently metabolised to glucose, the metabolic effects of the two monosaccharides are not the same. When considering the possible metabolic effects of lactose hydrolysate syrup the characteristic differences between galactose, glucose and lactose must be acknowledged.

Chronically elevated serum galactose levels are of potential danger to man, as demonstrated in clinical galactosaemia. Although ingestion of galactose results in high blood galactose levels, an effect which is galactose dose related, ingestion of glucose with galactose results in much lower blood galactose levels. In lactose hydrolysate syrup, galactose is accompanied by an equivalent amount of glucose and so high blood galactose levels would not be encountered. The adverse effect of alcohol on serum galactose levels after ingestion of galactose must, however, be remembered.

Many studies on the toxic effect of galactose are carried out in rats—animals which have a lower tolerance of galactose than man, and in which the modifying effect of glucose on blood galactose levels is not present. The effect of the lactose hydrolysate syrups on the fasting serum triglyceride concentration in female baboon remains an interesting point for further research.

## REFERENCES

1. HARJU, M. and KREULA, M. (1980). In: *Carbohydrate Sweeteners in Food and Nutrition*, P. Koivistoinen and L. Hyvonen (Eds.),. Academic Press, London, 233–42.
2. HUNT, J. N. (1961). *Gastroent.*, **41**, 49.
3. ELIAS, E., GIBSON, G. J., GREENWOOD, L. F., HUNT, J. N. and TRIPP, J. H. (1968). *J. Physiol.*, **194**, 317.
4. MALLINSON, C. N. (1968). *Gut*, **9**, 737.
5. NEWCOMER, A. D. and MCGILL, D. B. (1966). *Gastroent.*, **50**, 340.
6. MCMICHAEL, H. B., WEBB, J. and DAWSON, A. M. (1967). *Clin. Sci.*, **33**, 135.
7. MCMICHAEL, H. B. (1972). *Acta Hepato-Gastroenterologia*, **19**, 281.
8. GRAY, G. M. and SANTIAGO, N. A. (1966). *Gastroent.*, **51**, 489.
9. GRAY, G. M. and INGELFINGER, F. J. (1966). *J. Clin. Invest.*, **45**, 388.
10. COOK, G. C. (1973). *Clin. Sci.*, **44**, 425.
11. MACDONALD, I. and TURNER, L. J. (1968). *Lancet*, **1**, 841.
12. HOLZEL, A., SCHWARTZ, V. and SUTCLIFFE, K. W. (1959). *Lancet*, **1**, 1126.
13. DAHLQVIST, A., HAMMOND, J. B., CRANE, R. K., DUNDAY, J. V. and LITTMAN, A. (1963). *Gastroent.*, **45**, 488.
14. CUATRECASAS, P., LOCKWOOD, D. H. and CALDWELL, J. R. (1965). *Lancet*, **1**, 14.
15. COOK, G. C. and KAJUBI, S. K. (1966). *Lancet*, **1**, 725.
16. BAYLESS, T. M. and ROSENWEIG, N. S. (1966). *J. Amer. Med. Assoc.*, **197**(12), 966.
17. KRETCHMER, N., HURWITZ, R., RANSOME-KUTI, O., DUNGY, C. and AKKIJA, W. (1971). *Lancet*, **2**, 392.
18. ROSENSWEIG, N, S. (1981). *Gastroent.*, **60**, 464.
19. JOHNSON, J. D., KRETCHMER, N. and SIMOONS, F. J. (1974). *Adv. Paediatrics*, **21**, 197.
20. LAFAYETTE, P. and MENDEL, B. (1907). *Amer. J. Physiol.*, **20**, 81.
21. SUNSHINE, P. and KRETCHMER, N. (1964). *Science*, **144**, 850.
22. SIMOONS, F. J. (1967). *Amer. J. Dig. Dis.*, **14**, 819.
23. SIMOONS, F. J. (1970). *Amer. J. Dig. Dis.*, **15**, 695.
24. KRETCHMER, N. (1972). *Sci. Amer.*, **227**, 71.
25. KRETCHMER, N. (1976). In: *Paediatric Implications for Some Adult Disorders*, D. Baltrop (Ed.). Unigate Pediatric Workshops no. 4, Fellowship of Postgraduate Medicine, London, 65–70.
26. ADERS PLIMMER, R. H. (1906). *J. Physiol.*, **35**, 20.

27. HELSKOV, N. S. (1951). *Acta Physiol. Scand.*, **24**, 84.
28. ALVAREZ, A. and SAS, J. (1961). *Nature*, **190**, 826.
29. KOLDOVSKY, O. and CHYTIL, F. (1965). *Biochem. J.*, **94**, 266.
30. ROSENSWEIG, N. S. and HERMAN, R. H. (1969). *Amer. J. Clin. Nutr.*, **22**, 99.
31. DUNPHY, J. V., LITTMAN, A., HAMMOND, J. B., FORSTNER, G., DAHLQVIST, A. and CRANE, R. K. (1965). *Gastroent.*, **49**, 12.
32. NEWCOMER, A. D. and MCGILL, D. B. (1967). *Gastroent.*, **53**, 881.
33. PAIGE, D. M., BAYLESS, T. M., FERRY, G. D. and GRAHAM, G. G. (1971). *John Hopkins Med. J.*, **129**, 163.
34. PAIGE, D. M., LEONARDO, E., CORDANO, A., NAKASHIMA, J., ADRIAZEN, B. and GRAHAM, G. G. (1972). *Amer. J. Clin. Nutr.*, **25**, 297.
35. LESLIE, J., MACLEAN, W. C. JR. and GRAHAM, G. G. (1979). *Amer. J. Clin. Nutr.*, **32**, 971.
36. REDDY, V. and PERSHAD, J. (1972). *Amer. J. Clin. Nutr.*, **25**, 114.
37. JONES, D. V., LATHAM, M. C., KOSIKOWSKI, F. V. and WOODWARD, G. (1976). *Amer. J. Clin. Nutr.*, **29**, 633.
38. CORI, C. F. (1925). *J. Biol. Chem.*, **66**, 691.
39. GROEN, J. (1937). *J. Clin. Invest.*, **16**, 245.
40. CRANE, R. K. (1960). *Physiol. Rev.*, **40**, 789.
41. CRANE, R. K. (1977). In: *International Review of Physiol.*, **12**, R. K. Crane (Ed.). University Park Press, Baltimore, 325–65.
42. MCMICHAEL, H. B. (1973). *Gut*, **14**, 428.
43. DEBNAM, E. S. and LEVIN, R. J. (1976). *Gut*, **17**, 92.
44. LINDQVIST, B. and MEEUWISSE, G. W. (1962). *Acta Paediatr. (Uppsala)*, **51**, 674.
45. SCHNEIDER, A. J., KINTER, W. B. and STIRLING, C. E. (1966). *New Engl. J. Med.*, **274**, 305.
46. WIMBERLEY, P. D., HARRIES, J. T. and BURGESS, E. A. (1973). *Proc. R. Soc. Med.*, **67**, 755.
47. HOLDSWORTH, C. D. and DAWSON, A. M. (1964). *Clin. Sci.*, **27**, 371.
48. COOK, G. C. (1977). *Scand. J. Gastroent.*, **12**, 733.
49. SCHEDL, H. P. and CLIFTON, J. A. (1961). *J. Clin. Invest.*, **40**, 1079.
50. FINCH, L. R. and HIRD, F. J. R. (1960). *Biochim. Biophys. Acta*, **43**, 278.
51. COOMBE, N. B. and SMITH, R. H. (1973). *Brit. J. Nutr.*, **30**, 331.
52. CAHILL, G. F., OWEN, O. E. and FELIG, P. (1968). *The Physiologist*, **11**, 97.
53. LELOIR, L. F. (1951). *Arch. Biochem. Biophys.*, **33**, 186.
54. HERMAN, R. H. and ZAKIM, D. (1968). *Amer. J. Clin. Nutr.*, **21**, 127.
55. COHN, R. M. and SEGAL, S. (1973). *Metab.*, **22**(4), 627.
56. HANSEN, R. G. and GITZELMAN (1975). *ACS Symposium Series*, **15**, 100.
57. SOWDEN, J. (1957). In: *The Carbohydrates—Chemistry, Biochemistry, Physiology*, W. Pigman (Ed.). Academic Press Inc., New York, 88.
58. KALCKAR, H. M. (1965). *Science*, **150**, 305.
59. ISSELBACHER, K. J. (1957). *Science*, **126**, 652.
60. CASTRO, A., SCOTT, J. P., GRETTIE, D. P., MACFARLANE, D. and BAILEY, R. E. (1970). *Diabetes*, **19**, 842.
61. CHRISTENSEN, N. J., ORSKOV, H. and HANSEN, A. P. (1972). *Acta Med. Scand.*, **192**, 337.

62. JOURDAN, M. H. (1972). *Guy's Hosp. Rep.*, **121**(2, 3), 155.
63. MACDONALD, I., KEYSER, A. and PACY, D. (1978). *Amer. J. Clin. Nutr.*, **31**, 1305.
64. KING, E. J. and AITKEN, R. S. (1940). *Lancet*, **2**, 543.
65. COLCHER, H., PATEK, A. J. and KENDALL, F. E. (1946). *J. Clin. Invest.*, **25**, 768.
66. STENSTAM, T. (1946). *Acta Med. Scand.* (*Suppl.*), **177**.
67. BERNSTEIN, L. M. (1960). *Gastroent.*, **39**, 293.
68. TYGSTRUP, N. (1964). *Acta Med. Scand.*, **175**, 281.
69. ROMMEL, K., BERNT, E. and SCHMITZ, F. (1968). *Klin. Wschr.*, **46**, 936.
70. TENGSTRÖM, B. (1968). *Scand. J. Clin. Lab. Invest.*, **21**, 321.
71. ROE, J. H. and SCHWARTZMANN, A. S. (1932). *J. Biol. Chem.*, **96**, 717.
72. LEVINE, R., GOLDSTEIN, M. S., HUDDLESTUN, B. and KLEIN, S. P. (1950). *Amer. J. Physiol.*, **163**, 70.
73. WICK, A. N. and DRURY, D. R. (1953). *Amer. J. Physiol.*, **173**, 229.
74. POZZA, G. and GHIDONI, A. (1962). *Clinica Chimica Acta*, **7**, 55.
75. VEGA, F. V. and KONO, T. (1978). *Biochim. Biophys. Acta*, **512**, 221.
76. WILLIAMS, C. A., PHILLIPS, T. and MACDONALD, I. Accepted for publication in *Metabolism*.
77. MACDONALD, I., PHILLIPS, T. and KEYSER, A. Unpublished data.
78. HARDING, V. J. and GRANT, C. A. (1933). *J. Biol. Chem.*, **99**, 629.
79. CARPENTER, T. M. (1937). *J. Nutr.*, **13**, 583.
80. WILLIAMS, C. A. (1981). Some of the Metabolic Effects Associated with the Acute and Chronic Ingestion of Galactose in Man and Animals, PhD thesis, University of London.
81. CORI, C. F. (1926). *Proc. Soc. Exptl. Biol. Med.*, **23**, 290.
82. DEUEL, H. J. (1936). *Physiol. Rev.*, **16**, 173.
83. MORGAN, L. M., WRIGHT, J. W. and MARKS, V. (1979). *Diabetologica*, **16**, 235.
84. DORMANDY, T. L. (1959). *Lancet*, **2**, 269.
85. GANDA, O. P., SOELDNER, J. S., GLEASON, R. E., CLEATOR, I. G. and REYNOLDS, C. (1979). *J. Clin. Endocrinol. Metab.*, **49**(4), 616.
86. SEGAL, S. and BLAIR, A. (1961). *J. Clin. Invest.*, **40**, 2016.
87. SHIMA, K., KURODA, K., MATSUYAMA, T., TARUI, S. and NIHIKAWA, M. (1972). *Proc. Soc. Exp. Biol. Med.*, **139**, 1042.
88. AMBRUS, J. L., AMBRUS, C. M., SHIELDS, R., MINK, B. and CLEVELAND, C. (1976). *J. Med.*, **7**, 6.
89. SAMOLS, E. and DORMANDY, T. L. (1963). *Lancet*, **1**, 478.
90. GITZELMAN, R. and ILLIG, R. (1969). *Diabetologia*, **5**, 143.
91. ROYLE, G., KETTLEWELL, M. G. W., ILIC, V. and WILLIAMSON, D. H. (1978). *Clin. Sci. Mol. Med.*, **54**, 107.
92. GRODSKY, G. M., BATTS, A. A., RENNETT, L. L., VCELLA, C., MCWILLIAMS, N. B. and SMITH, D. F. (1963). *Amer. J. Physiol.*, **205**, 638.
93. SUSSMAN, K. E., VAUGHAN, G. D. and TIMMER, R. F. (1966). *Metab.*, **15**, 466.
94. GOETZ, F. C., MANEY, J. W. and GREENBERG, B. Z. (1967). *J. Lab. Clin. Med.*, **69**, 537.
95. MALAISSE, W., MALAISSE-LAGAE, F. and WRIGHT, P. H. (1967). *Endocrinol.*, **80**, 99.

48  CELIA A. WILLIAMS

96. DEUEL, H. J. JR. (1927). *J. Biol. Chem.*, **75**, 367.
97. MACDONALD, I. and RUSSELL, C. Unpublished data.
98. DUNCAN, D. L. (1955). *Nutrition Abstr. Rev.*, **25**, 309.
99. ALI, R. and EVANS, J. L. (1973). *J. Agric. Univ. of Puerto Rico*, **57**, 149.
100. MCBEAN, L. D. and SPECKMANN, E. W. (1974). *Amer. J. Clin. Nutr.*, **27**, 603.
101. FOOD AND AGRICULTURE ORGANIZATION OF THE UNITED NATIONS (1980). In: *Food and Nutrition*, Paper 15, 46.
102. FOURNIER, P. L., DUPUIS, Y., SUSBIEVE, M., ALLEZ, M. and TARDY, N. (1955). *J. Physiol.* (*Paris*), **43**, 339.
103. WASSERMAN, R. H. and COMAR, C. L. (1959). *Proc. Soc. Exp. Biol. Med.*, **101**, 314.
104. VAUGHAN, O. W. and FILER, L. J., JR. (1960). *J. Nutr.*, **71**, 10.
105. CONDON, J. R., NASSIM, J. R., MILLARD, F. J. C., HILBE, A. and STAINTHORPE, E. M. (1970). *Lancet*, **1**, 1027.
106. BUSHNELL, P. J. and DELUCA, H. F. (1981). *Science*, **211**, 61.
107. ELIA, M., OPPENHEIM, W. L., SMITH, R., ILIC, V. and WILLIAMSON, D. H. (1979). *Clin. Sci.*, **57**, 249.
108. PHILLIPS. T., MACDONALD, I. and KEYSER, A. (1978). *Proc. Nutr. Soc.*, **37**, 24A.
109. CORI, C. F. and CORI, G. T. (1926). *J. Biol. Chem.*, **70**, 577.
110. DEUEL, H. J. JR., MACKAY, E. M., JEWEL, P. W., GULICK, M. and GRUNEWALD, C. F. (1933). *J. Biol. Chem.*, **101**, 301.
111. CORI, C. F. and CORI, G. T. (1934). *Ann. Rev. Biochem.*, **3**, 151.
112. HARDING, V. J., GRANT, G. A. and GLAISTER, D. (1934). *Biochem. J.*, **28**, 257.
113.| WAGNER, F. R. (1914). *Zeitschr. f. klin. Med.*, **80**, 174.
114. SALASPURO, M. P. and KESANIEMI, Y. A. (1973). *Scand. J. Gastroent.*, **8**, 681.
115. BAUER, R. and WOŻASEK, O. (1934). *Wein Klin. Wchnschr.*, 995.
116. ISSELBACHER, K. J. and KRANE, S. M. (1961). *J. Biol. Chem.*, **236**, 2394.
117. HERMAN, R. H. (1974). In: *Sugars in Nutrition*, H. L. Sipple and K. W. McNutt (Eds.). Academic Press, New York, 146–72.
118. WILLIAMS, C. A. and MACDONALD, I. (1982). *Human Nutr.: Clin. Nutr.*, **36C**, 149.
119. KRITCHEVSKY, D. *et al.* (1980). *Amer. J. Clin. Nut.*, **33**, 1869.
120. KRITCHEVSKY, D., SHAPIRO, I. L. and WERTHESSEN, N. T. (1962). *Biochim. Biophys. Acta*, **65**, 556.
121. SHAPIRO, I. L. and KRITCHEVSKY, D. (1963). In: *Proceedings of the First International Symposium on the Baboon and its Use as an Experimental Animal, San Antonia*. H. Vartborg (Ed.). University of Texas Press, Austin, 339–52.
122. GRESHAM, G. A. and HOWARD, A. N. (1969). In: *Experimental Medicine and Surgery in Primates*, E. I. Goldsmith and J. Moor Jankowski (Eds.); *Annals of the New York Academy of Sciences*, **162** (Art 1), 99.
123. MANNING, G. W. (1942). *Amer. Heart J.*, **23**, 719..
124. MCGILL, H. C. JR., STRONG, J. P., HOLMAN, R. L. and WERTHESSEN, N. T. (1960). *Circulation Res.*, **8**, 670.
125. LARKSON, T. B. (1965). In: *Comparative Atherosclerosis*, J. C. Roberts, Jr. and R. Strauss (Eds.). Harper and Row, New York, 211–14.
126. ANTOIS, A. and BERSOHN, I. (1961). *Lancet*, **1**, 3.
127. BIERMAN, E. L. and HAMLIN, J. T. (1961). *Diabetes*, **10**, 432.

128. CONNOR, W. E. (1979). In: *Biochemistry of Atherosclerosis*, A. M. Scanu (Ed.). Marcel Dekker Inc., New York and Basel, 371–418.
129. ALBRINK, M. J. and MAN, E. B. (1959). *Arch. Intern. Med.*, **103**, 4.
130. CROWLEY, L. V. (1971). *Clin. Chem.*, **17**, 206.
131. CARLSON, L. A. and BÖTTIGER, L. E. (1972). *Lancet*, **1**, 865.
132. MACDONALD, I. and BRAITHWAITE, D. M. (1964). *Clin. Sci.*, **27**, 23.
133. MACDONALD, I. (1965). *Clin. Sci.*, **29**, 193.
134. BEVERIDGE, J. M. R., JAGANNATHAN, S. N. and CONNELL, W. F. (1964). *Can. J. Biochem.*, **42**, 999.
135. MACDONALD, I. and BRAITHWAITE, D. M. (1964). *Clin. Sci.*, **27**, 23.
136. MACDONALD, I. (1965). *Amer. J. Clin. Nutr.*, **16**, 458.
137. KALCKAR, H. M., ANDERSON, E. P. and ISSELBACHER, K. J. (1956). *Biochim. Biophys. Acta*, **20**, 262.
138. SCHWARZ, V., GOLDBERG, L., KOMROWER, G. M. and HOLZEL, A. (1956). *Biochem. J.*, **62**, 34.
139. ANDERSON, E., KALCKAR, H. and ISSELBACHER, K. (1957). *Science*, **125**, 113.
140. GITZELMAN, R. (1967). *Pediatr. Res.*, **1**, 14.
141. SCHWARZ, V. (1975). *Biochem. Soc. Trans.*, **8**, 234.
142. RENNERT, O. M. (1977). *Ann. Clin. Lab. Sci.*, **7**, 443.
143. GITZELMAN, R. and HANSEN, R. G. (1980). In: *Inherited Disorders of Carbohydrate Metabolism*, D. Burnman, J. B. Holton and C. A. Pennock (Eds.). MTP Press, Lancaster, 61–101.
144. MITCHELL, H. S. and DODGE, W. M. (1935). *J. Nutr.*, **9**, 37.
145. KORC, I. (1961). *Arch. Biochem. Biophys.*, **94**, 196.
146. SEGAL, S. and BERSTEIN, H. (1963). *J. Pediat.*, **62**, 363.
147. PATTERSON, J. W. and PATTERSON, M. E. (1965). *Proc. Soc. Exp. Biol. Med.*, **118**, 324.
148. FOURNIER, D. J. and PATTERSON, J. W. (1970). *Proc. Soc. Exp. Biol. Med.*, **135**, 377.
149. RICHTER, C. P. and DUKE, J. R. (1970). *Science*, **168**, 1372.
150. KEIDING, S. and MELLEMGAARD, L. (1972). *Acta. Ophthal.* (*KBL*), **50**, 174.
151. BANNON, S. L., HIGGINBOTTOM, R. M., MCCONNELL, J. M. and KAAN, H. W. (1945). *Arch. Ophth.*, **33**, 224.
152. TYGSTRUP, N., SCHMIDT, A. and THIEDEN, H. I. D. (1971). *Scand. J. Clin. Lab. Invest.*, **28**, 27.
153. NEWSTEAD, C. G. (1979). *Proc. Nutr. Soc.*, **38**(2), 38A.
154. SRIVASTA, S. K. and BEUTLER, E. (1970). *Experientia*, **26**, 250.
155. BHAT, K. S. (1974). *Nutr. Metab.*, **16**(2), 111.
156. PATTERSON, J. W. and BUNTING, K. W. (1964). *Proc. Soc. Exp. Biol. Med.*, **115**, 1156.
157. VAN HEYNINGEN, R. (1971). *Exp. Eye Res.*, **11**, 415.
158. REDDY, V. N., SCHWASS, D., CHAKRAPANI, B. and LIM, C. P. (1976). *Exp. Eye Res.*, **23**, 483.
159. UNAKAR, N. J., GENYEA, C., REDDAN, J. R. and REDDY, V. N. (1978). *Exp. Eye Res.*, **26**, 123.
160. VAN HEYNINGEN, R. (1959). *Nature*, **184**, 194.
161. KINOSHITA, J. H., MEROLA, L. O., SATOH, K. and DIKMAK, E. (1962). *Nature*, **194**, 1085.

162. KINOSHITA, J. H., MEROLA, L. O. and DIKMAK, E. (1962). *Exp. Eye Res.*, **1**, 405.
163. KINOSHITA, J. H. (1965). *Invest. Ophthalmol.*, **4**, 786.
164. BUNCE, G. E. (1979). *Nutrition Reviews*, **37**, 337.
165. SPATZ, M. and SEGAL, S. (1965). *J. Pediatr.*, **67**, 438.
166. HAWORTH, J. C., FORD, J. D. and YOUNOSZAI, M. K. (1969). *Pediatr. Res.*, **3**, 441.

*Chapter* 3

# NUTRITIVE SUCROSE SUBSTITUTES AND DENTAL HEALTH

T. H. GRENBY

*Department of Oral Medicine and Pathology,*
*Guy's Hospital, London, UK*

*SUMMARY*

*The effects of the current level of sucrose consumption on dental health and obesity are outlined. Alternative nutritive sweeteners are classified into four main groups: sugars; starch hydrolysates; sugar alcohols; and other products. Information published on the sweeteners is given in the order: source and availability; composition; regulatory status; application; general nutritional considerations and physiological effects; and influence on dental health, with data from clinical trials, laboratory animal experiments and studies* in vitro.

## 1. INTRODUCTION

Sucrose is a major item of diet in many developed countries, in demand particularly for its sweet taste, its preservative properties, and the desirable physical characteristics it imparts in both food processing and consumption. In Britain about 19 % of the nation's total energy requirements are supplied by sugars. Approximately 16 % of this is provided by sucrose, with the remaining 3 % mostly from glucose and glucose products, although this figure is increasing slowly year by year. Out of a total daily sugar usage of 120 g per person, about 28 % or 34 g is in the form of confectionery. The figures for chocolate and sugar confectionery consumption in 1978, taking an average for the whole population, were 18 and 16 g per day, respectively,[1] although these figures fell slightly in 1979–1981 to 17 and 14 g per day, respectively.[2] The Ministry of Agriculture, Fisheries and Food[3] estimated that in 1973 sweets and chocolate alone provided almost 5 % of the average daily energy intake. Apart from that used as pure sugar for sweetening tea and coffee, for the preparation of foods in the home, and sprinkling on breakfast cereals, the remainder of the sugar is used in the manufacture of a wide range of foods and in brewing.

In the last 130 years sucrose has risen from being a minor food item to a major constituent of the diet. In the UK in 1978 expenditure on confectionery was one-and-a-half times greater than on bread. There are mixed views in the scientific and medical community about this state of affairs. At one extreme, the high sucrose intake in westernised countries has been indicted as a factor in the aetiology of a whole range of disorders from diabetes and heart and arterial disease to obesity, eye and skin diseases, and dental caries.[4–6] At the other end of the scale are nutritionists with the more moderate view that the origins of most of these disorders are multifactorial, and that clear evidence incriminating sucrose consumption alone as the causative agent is lacking. The epidemiological evidence is notably inconclusive. Nevertheless there is support for the view that excessive and imbalanced consumption of high-sucrose foods is a contributory factor in the onset of both obesity and dental caries.

Obesity is probably the most serious nutritional disorder facing developed countries.[7] Surveys have shown that about 20 % of infants and about 25 % of the total population may be classed as obese (for more detailed data see Reference 8). This is a major problem because of the many health risks attendant upon obesity, including diabetes, damage to weight-bearing joints, complications after surgery and in childbirth, reduced respiratory function and psychological disturbances.[9]

Obesity is not the consequence of excessive carbohydrate consumption alone but can be produced by immoderate intake of fats, proteins and carbohydrates over and above an individual's normal requirements and his ability to eliminate or metabolise them. The reason that high-sucrose foods are so often blamed is that they tend to be eaten in addition to an adequate diet, especially as between-meals snacks, and their high palatability and convenient packaging help to make them readily acceptable even after nutritional needs and appetite have been largely satisfied. It is not clear whether the replacement of sucrose in such foods will bring about any significant nutritional benefits. Only about 25 % of the total daily sucrose consumption in the UK is in the form of confectionery, with a further 4–5 %

in biscuits eaten in the home. If new products can be formulated in which sucrose is replaced by non-caloric ingredients this might help to reduce total calorie intake. Replacing sucrose by an iso-caloric substance, i.e. one providing the same number of calories as the same weight of sucrose, will not solve the obesity problem unless eating patterns are also changed. However, it has been shown in the last 20 years or so that carbohydrates vary in their capacity to promote the laying down of fat in the body, with sucrose at or near the top of the list of common lipogenic food carbohydrates.[10-15] In most of these studies, blood lipid levels or fat deposition were greater on sucrose than on glucose or glucose-syrup-containing regimes. There is thus some basis for investigating the replacement of sucrose in foods by other carbohydrate substances as a step to limiting body-fat deposition.

The other area in which the aetiological role of sucrose is well established is dental disease. Although it can be argued that many features of our diet and way of life contribute to the very widespread incidence of dental caries, there is an overwhelmingly substantial body of clinical and scientific evidence linking it to ill-advised sucrose-eating habits (see, for example, References 16, 17 and 18). More than 99 % of the population of the UK suffer or have suffered from dental caries. In a survey[19] carried out in England and Wales in 1968 about 40 % of the remaining teeth in adults over 16 years of age had been attacked by caries; the corresponding figure for Scotland in 1972 was 61 %.[20] By the age of five about 75 % of all children in England and Wales suffer from dental caries, and more than 33 % of the teeth of 15-year-olds are already decayed, missing or filled as a consequence of dental caries.

Sucrose is involved in the dental decay process in two ways. First, it is an energy-giving substrate for many of the micro-organisms that flourish on and around the teeth. Under conditions of limited oxygen availability the end-products of the metabolism of sucrose include organic acids, particularly lactic acid, which can attack tooth mineral at certain sites, for example, between the teeth and deep in the molar fissures, where the acids can accumulate. If sucrose-containing foods are taken into the mouth frequently, wave after wave of acid is produced, favouring demineralisation of the tooth enamel and the development of a carious lesion. Sucrose is not the only sugar which can act as a substrate for acid production—glucose, for example, is used by the oral micro-organisms in a similar manner.

The properties which make sucrose more cariogenic than the other common sugars are believed to be a consequence of its role as a substrate for the formation of dental plaque, in which oral micro-organisms are organised into colonies embedded in a gelatinous matrix which adheres to the tooth surface. Plaque tends to build up in inaccessible areas and crevices by multiplication of the bacterial colonies and the formation of more matrix, which contains polysaccharides (glycans) synthesised by the bacteria themselves. Certain bacteria can use sucrose as a substrate for this process without first hydrolysing it to glucose and fructose, utilising the energy of the glycosidic bond to synthesise the complex glycan polymers directly. The matrix not only acts as a 'sponge' holding sucrose, bacteria and their acid products close up against the tooth surface, but may also be used later, along with intracellular polysaccharide, as an energy source by the acidogenic bacteria.

Dental plaque also plays a part in the onset of periodontal disease in addition to caries. Surveys have shown that caries and periodontal disease exist together in children. Of children with caries, 78 % of 5-year-olds and 67 % of 12-year-olds had gingivitis. In another study, 89 % of 11-year-olds and 69 % of 17-year-olds had gingivitis.[21] In adults in England and Wales, gum disease was recorded around 25 % of the teeth in the 16–34 age group, and 50 % in the group aged 35 and over.[19] If plaque is allowed to build-up at the gingival margin of the teeth and to remain in contact with the gingiva, bacteria can penetrate and infect the gingival tissues, promoting inflammatory changes and leading to chronic gingivitis. If the plaque is allowed to remain in position at these sites it will gradually harden to calculus. If the calculus is allowed to extend towards the root of the tooth it can force the supporting tissues away, loosening it and eventually leading to the loss of the tooth.

To bring about an improvement in dental health any substitute sweetener ideally should be a poorer substrate than sucrose for the bacterial production of acids and plaque polysaccharides. Of course, many other factors are involved in the choice of sucrose substitutes, including safety, cost, physiological effects other than dental, behaviour in food processing, and effects on the characteristics and acceptability of foods containing them, and any information on these will be dealt with below as each type of substitute is considered.

## 2. SUCROSE SUBSTITUTES

Sweet-tasting alternatives to sucrose available to the food manufacturer fall into two categories: caloric and non-caloric.

Non-caloric sweeteners include a range of natural products currently under intensive scrutiny and some synthetic chemicals such as saccharin and cyclamate, but they are not to be considered here. One of their main uses is in sweetening drinks, for which 'bulking' properties are not required, but another important function that should not be overlooked is their usefulness when blended with caloric sucrose substitutes less sweet than sucrose, to bring the level of sweetness up to the level of the original sucrose formulation.

Caloric sweeteners are carbohydrates or carbohydrate derivatives or mixtures of products from the chemical or enzymic treatment of carbohydrate materials, mainly of plant origin. Animal food products contain very little carbohydrate. The list of materials available is in a state of active expansion. This is prompted not only by the growing recognition of excessive sucrose intake as a health hazard but also by commercial and economic considerations. In the last few years suppliers of raw materials to the food-manufacturing industry have developed methods for processing starches and other polymeric plant products to provide a whole range of sweet food ingredients to meet varying requirements. Their efforts were given impetus by the shortage and consequent high price of sugar in 1974–5. An advantage of using starch as a starting material is that the manufacturing processes can be adapted to encompass starch from different sources including potatoes and various cereals, depending on their availability. The same considerations apply in choosing the source of xylans for the production of xylitol.

One point often overlooked by food reformers and others with a strong desire to reduce the sucrose content of the diet is that sucrose is not used in foods to provide sweetness alone. It has, in addition, a whole range of useful properties, particularly as a preservative, in providing bulk, improving physical properties and enhancing fruit flavours, etc.[22,23] The confectionery and chocolate industry is based on an understanding of sucrose technology built up over many years,[24] and it is not practicable simply to remove sugar from a well-accepted item of food and replace it overnight with another sweetener that may have very different properties necessitating major alterations in manufacturing processes. Consumer acceptance of the new product and its price relative to the original sucrose product also have to be taken into account.

The caloric sweeteners that show some potential at present as alternatives to sucrose can be classified in Sections 3–6 as follows:

3. *Sugars*
    3.1. Glucose
    3.2. Fructose
    3.3. Maltose
    3.4. Sorbose

4. *Starch hydrolysates*
    4.1. Glucose syrups
    4.2. High-fructose syrups
    4.3. Hydrogenated glucose syrups (Lycasin ®*)

5. *Sugar alcohols*
    5.1. Sorbitol
    5.2. Mannitol
    5.3. Maltitol
    5.4. Isomaltitol
    5.5. Xylitol
    5.6. Palatinit ®†
    5.7. Lactitol

6. *Other products*
    6.1. Coupling sugar

Information to hand on these products will be dealt with in the following
order:

1. Source and availability;
2. Composition of the product;
3. Regulatory position—if known;
4. Application—types of food in which it may be incorporated, and
   any known technological advantages or disadvantages;
5. General nutritional considerations and physiological effects;
6. Influence on dental health—data available from clinical, laboratory animal and *in vitro* studies.

Clinical studies obviously give the best guide to the value of individual
sucrose substitutes in dental health, but they can seldom be done for reasons
of cost, ethics and the amount of organisation and supervision they require.
Still, in the field of human data, epidemiological studies can also be of some
value, but the findings often relate to a whole diet or eating pattern and not
to the substitution of one specific item of food by another. This is where
animal experiments can give more precise information. Accurately
formulated diets containing the carbohydrates or substitutes under test can
be fed to matched groups of animals in controlled amounts, and their
dental, nutritional and toxicological effects observed. Strains of laboratory
animal bred for uniform susceptibility to dental caries are available. Apart

* Lycasin ® is a registered trade mark of Roquette Frères, France.
† Palatinit ® is a registered trade mark of Süddeutsche Zucker AG, West Germany.

from such studies *in vivo*, a range of tests *in vitro* may be used to assess the role of different foods in inducing or suppressing caries. These may involve the action of oral micro-organisms on the foods, measured in terms of acid or polysaccharide production, and their influence on prepared samples of dental enamel. Each of these approaches has limitations, but they all have a part to play in building-up a complete picture of the usefulness of a sugar substitute. The difficulty lies in meeting regulatory requirements and interpreting the data from different sources to decide what application each sweetener could have in the food industry.

## 3.    SUGARS AS ALTERNATIVES TO SUCROSE

### 3.1.    D-Glucose

D-Glucose is the monosaccharide—formula $C_6H_{12}O_6$—usually known as dextrose. The term 'glucose' is sometimes used loosely in the food industry to embrace hydrolysed starch or glucose syrups, which are dealt with in Section 4.1. D-Glucose is cheap and readily available, and has a mean sweetness, relative to sucrose, of about 0·7.[25] Therefore, in certain food applications, additional sweetening from non-caloric agents is required to bring the sweetness up to the level provided by the sucrose being replaced.

Figures for the consumption of glucose alone, as distinct from glucose syrups, are difficult to obtain for the UK, but in the USA in 1974 glucose accounted for 6 % of the total sugar intake, i.e. about 12 g per person per day.[26]

Glucose occupies a central position in the body's metabolic processes. When dietary carbohydrates enter the body they are converted in the main to glucose which is then taken up by the bloodstream for transport around the body. Glucose is used: (1) in an oxidative process to provide energy for both muscular work and the maintenance of local metabolism; (2) to synthesise storage polysaccharide (glycogen) in the liver and muscles; and (3) as a starting material in the synthesis of other body constituents including fats and proteins. These processes are described in many standard biochemical textbooks, but see also Reference 27.

Glucose is on the GRAS (generally recognised as safe) list of the US Food and Drug Administration. As an energy source, it has been equated customarily with sucrose, although evidence is now accumulating that it is less lipogenic than sucrose (see Section 1 for references). Glucose is used by food manufacturers primarily as a sweetener, but its property of caramelisation is also of value. It cannot be used in foods as a straightforward one-to-one replacement for sucrose because there are certain differences in physical properties between the two sugars, e.g. solubility, melting point and crystallinity.

There is not much precise information in the literature on the cariogenicity of glucose. Very little clinical work has been carried out on glucose as opposed to glucose-syrup products, but in one trial the replacement of sucrose in the diet by a low-calorie sweetener compounded of glucose and saccharin was tested.[28] Dental plaque formation and composition were examined in young adults who had used either sucrose or the low-calorie sweetener whenever possible in their food and drink over three-day periods. The results showed that the amount of plaque on the teeth was reduced after the period on the low-calorie sweetener and there was a good correlation between the individuals' plaque scores and their dental experience. This is indirect evidence relating to the cariogenicity of glucose. In several other studies a correlation has been found between the tendency to form dental plaque and caries activity. Considering the two properties of sugars believed to play a part in determining their cariogenicity (see Section 1), glucose shows no advantage over sucrose in terms of acid production—in fact when standard cultures of oral micro-organisms are incubated in media containing the two sugars, acid production is generally more rapid from glucose than from sucrose. With respect to plaque polysaccharide synthesis, however, sucrose is a better substrate than glucose (bacteriological aspects of the metabolism of sucrose in plaque have been reviewed by Cole[29]).

More direct information on the cariogenicity of glucose is available from animal experiments. Both sucrose and glucose have been used as dietary ingredients to induce dental caries in laboratory rodents, and some workers in the field say there appears to be little or no difference between them. This may be the case when the animals have a very low susceptibility to caries, but under conditions of increased sensitivity there is good evidence that glucose is less cariogenic than sucrose.[30–6] In one series of experiments, the reduction in mean caries scores brought about by replacing the 33 % of sucrose in a diet by glucose varies between 18 and 59 %.

The other possible dental benefit that might accrue from the use of glucose in foods in place of sucrose is an improvement in periodontal condition as a result of diminished dental plaque formation.

### 3.2. Fructose

Fructose is a component hexose, linked by a glycosidic bond to glucose in the disaccharide sucrose. Fructose (laevulose or 'fruit sugar') is also found in the free state in certain fruits and vegetables, honey, some animal tissues and blood. It is present, polymerised as inulin, in the tuberous roots of many Compositae, such as artichokes and dahlias. In normal diets, fructose is said to contribute 15 to 50 % of the total carbohydrate intake, but most of this is in combination as sucrose. Commercially, it can be produced from sucrose by inversion (hydrolysis) or from glucose by chemical or enzymic isomerisation,[37] but it is more expensive than either sucrose or glucose.

Unlike glucose and most of the other common hexoses, fructose is a keto-sugar and exists in solution as a mixture of pyranose and furanose forms. It also differs from the other common sugars in physical properties, e.g. it is more hygroscopic. Its properties and uses are considered in detail by Osberger and Linn.[38]

Nutritional and dental interests in fructose seem to have increased over the last few years since it has become available commercially, and various ideas have been put forward on how fructose could be used in foods and for special metabolic purposes. One factor that has counted in its favour is that it is 1·2–1·8 times as sweet as sucrose, so that where sweetening is the only function of sucrose it could be replaced, in theory, by a smaller amount of fructose.

Some possible applications of fructose in the food industry are reviewed by Pawan[39] and Osberger and Linn.[38] In particular, some authorities regard fructose as a suitable sweetening agent for diabetics, and it may find other uses in confectionery, bakery products and soft drinks.

There is a large amount of literature on the metabolism of fructose, although not all nutritionists agree on its suitability either for diabetics or as a replacement for a major part of the sucrose in the diet.[40] There is probably an active transport mechanism for the absorption of fructose in the alimentary tract, but it is absorbed more slowly than glucose. It is metabolised without the need for insulin, mainly in the liver, the kidney and the small intestine, via a specialised pathway, resulting in some lactate and pyruvate as breakdown products and also some glucose and glycogen.[39,41] Controversy over the suitability of fructose as a sucrose replacement centres around its property of elevating blood triglyceride levels compared with glucose[150] but, apart from this, a number of advantages of having fructose in the diet in certain circumstances are listed by Palm.[42]

The results of clinical dental trials of fructose among other caloric replacements for sucrose are summarised by Scheinin,[43] but fructose was tested in only two of them. The first was the well-known Turku clinical study in which 125 young Finns were divided into three groups and given foods containing sucrose, fructose or xylitol as the sole sugar. The sucrose group served as control. The other two groups received foods in which all the sucrose had been replaced by either fructose or xylitol.[44] The intention was to make the fructose foods as similar as possible to their normal sucrose-containing counterparts. Confectionery included chocolate bars and slabs with a fructose content of 28–30%, and various sorts of lozenges, pastilles and candies for which the exact fructose content is not quoted. Most fructose products were made by the simple replacement of sucrose, but sometimes the extra sweetness of fructose led to a reduction in the amount used. The overall consumption of sucrose and fructose was almost the same, however. According to Mäkinen and Scheinin,[44] the fructose products were generally considered equal to, and in some cases (jams, juices and pastries) superior to, the sucrose-containing products.

After two years of this almost complete replacement of sucrose by fructose a reduction in caries of 32% was reported, attributable entirely to a low increment rate of incipient lesions in the fructose group.[43] Scheinin et al.[45] quote the incidence of caries as 25% lower than in the sucrose group. Expressed another way, the mean increment in decayed, missing and filled tooth surface (DMFS index) was 7·2 in the sucrose group and 3·8 in the fructose group.[46] These increments were even smaller with xylitol. Many other aspects of the metabolism and physiological effects of fructose were examined in a series of 21 papers on the trials, and no health problems were encountered.[47]

Very few details are available about a more recent trial of fructose and glucose sweets in the diet of 70 three- to six-year-old children lasting one year ('Gustavsberg Study') in which a 39% reduction in caries was recorded by Frostell et al. (reported by Scheinin).[43]

In studies in animals, Gustafson et al.[31] observed that both sucrose and fructose produced more caries than glucose in hamsters, while others[34,35] found that sucrose generally gave rise to more caries than either glucose or fructose in rats. Later work in rats confirmed that sucrose was more cariogenic than glucose and fructose, and showed no consistent difference between the two monosaccharides.[36]

Very little evaluation of fructose in caries tests in vitro appears to have been done, but unpublished data on its behaviour as a substrate for acid-producing oral micro-organisms show that acid is produced less readily from fructose than from sucrose or glucose. This may be partly a question of adaptation—common oral micro-organisms may be adapted to utilise preferentially the two commonest dietary sugars, sucrose and glucose, and may only begin to metabolise fructose readily in the absence of sucrose and glucose over a period of time.

As in the case of glucose, current dental research opinion inclines to the view that fructose is less cariogenic than sucrose because it is a less suitable substrate for the synthesis of the polysaccharides that constitute the matrix of the dental plaque. Although the plaque is believed to contain mainly polymers of glucose (glycans), small amounts of levans or fructosans (fructose polymers) have also been detected, and from thermodynamic considerations it is thought likely that sucrose rather than fructose is the starting material for their synthesis. It has been suggested that fructose is less cariogenic than sucrose because the levans in the plaque are much more soluble than the glycans,[38] but this is probably not the whole explanation.

### 3.3. Maltose
Maltose is a reducing disaccharide which contains two glucose molecules linked by an $\alpha$-1,4-bond. It is found in nature in certain fruits, e.g. grapes, and as a hydrolysis product of starch. It has 0·3–0·6 times the sweetness of sucrose.[25] Maltose accounted for 2·7 % of the total sugars in the US diet in 1972.[48] There should be no regulatory objection to its unrestricted use in foods because: (1) it is a normal breakdown product of food starches; and (2) it is broken down solely to glucose, which can then enter the body's normal metabolic pathways.

Maltose is hydrolysed by the $\alpha$-glucosidase enzyme maltase in the villi of the small intestine, and the resulting glucose is actively and rapidly absorbed. A detailed description of the metabolic effects of maltose is given by Macdonald,[49] but most of the data relate to glucose syrups containing maltose plus a variety of other gluco-oligosaccharides, and there appears to be little specific information on the metabolism of maltose alone (the metabolism of the glucose syrups is described in Section 4.1). One advantage put forward for maltose and its polymers is that they have a lower osmotic pressure than monosaccharides and related polyols, and therefore show a reduced osmotic effect in the gut, leading to less risk of abdominal discomfort, anorexia and nausea. Like glucose, maltose has a smaller adverse influence on lipid metabolism than sucrose.

There is not much information available on maltose and dental health. No clinical trials appear to be on record, but a limited amount of animal work has been done. Shaw et al.[33] found that maltose in the diet of rats gave rise to as rapid a development of dental caries lesions as sucrose, but was responsible for less smooth-surface caries in one out of two experiments. Maltose also produced fewer signs of a periodontal syndrome than sucrose. The results of Green and Hartles[35] suggested that maltose was less cariogenic than fructose in albino rats.

Maltose was the chief sugar in some experimental non-sucrose sweet biscuits tested for cariogenic effects in rats.[50] After it had been found that sweet biscuits are among the most highly cariogenic of all the foods tested in the diet of laboratory rats,[51] various attempts at reformulation were made. In one of these, the 15 % sucrose in wheatmeal 'digestive' biscuits was replaced by a mixture of 5 % maltose (malt extract, 29 % maltose) and 1 % each of lactose, glucose and maltotriose, together with saccharin for added sweetness. The maltose-containing biscuits produced far less caries than the conventional digestives, with a mean reduction in caries score of approximately 66 % and a significant reduction in dental plaque deposits.

In earlier rat-feeding trials maltose seemed to produce more extracellular material in dental plaque than glucose.[33] This is in line with the view that these polysaccharide substances are synthesised by micro-organisms using the energy of the bond linking the two monosaccharide units in a disaccharide molecule, but not so easily from the monosaccharides alone.

### 3.4. L-Sorbose

L-Sorbose is a ketohexose sugar which is isomeric with fructose but belongs to the L- series as opposed to the more commonly occurring D- series. The principle of modern production is the fermentation (oxidation) of sorbitol by *Acetobacter suboxydans*.[52] The level of sweetness of sorbose is 0·8 relative to sucrose, but in confectionery and chocolate it may appear to be as sweet as sucrose. Unlike the polyols (Section 5), sorbose can undergo the normal sugar colour-producing changes of caramelisation and Maillard reaction in the presence of amino-acids under appropriate conditions.

Sorbose slows down the solidification of hard-boiled candy, and is not recommended for use in this type of product. It can be used in compressed candies for chewing gum, and it can also be incorporated in coatings for jellied candy, etc. It is said to be especially suitable in the standard chocolate-making procedure and enrobing where other sucrose substitutes may present technological problems—however, in milk chocolate it must be used together with a lactose-free dairy product.

Very little information is available on the regulatory status of L-sorbose, but it is said to have the 'zahnschonend' (safe for the teeth) characteristics applied to certain products in Switzerland as a result of a standard oral acid-production test devised by Mühlemann.[53,54] L-Sorbose is not yet on the market in the UK, and is not mentioned in the regulations on food constituents, but it fits the definition of carbohydrate-sweetening matter that may be used in sugar confectionery, although its status for diabetic claims and chocolate is more doubtful.

A limited amount of work on the metabolism of L-sorbose in the rat has been reported.[55] Incorporated at a level of 10% in the diet, some of the metabolic responses were comparable with those of D-glucose, but among several significant changes, caecum and liver weights were raised, and the authors concluded that the safety of L-sorbose requires very thorough studying before unrestricted use in human nutrition could be recommended.

Beereboom[56] notes that since L-sorbose is usually prepared by the oxidation of sorbitol it is likely to be expensive, and there is at present little evidence to support the non-cariogenic and low-caloric claims that have been made for it.[57] The non-cariogenic claim appears to be based on the finding that L-sorbose is either not or only slightly fermented by oral bacteria.[52] This is indirect evidence, but no reports of clinical or animal trials have yet appeared in the literature.

## 4.  STARCH PRODUCTS TO REPLACE SUCROSE

### 4.1.  Glucose Syrups (sometimes called 'corn syrups')

There is a whole range of glucose syrups, essentially starch hydrolysates, manufactured by several large companies. The starch may be derived from a variety of plant sources but in practice the usual starting materials are maize, wheat or potato starch. The long chains of glucose units constituting the starch molecules are subjected to attack by acids or enzymes under controlled conditions, producing mixtures of glucose, maltose and higher oligosaccharides. The composition of the final mixture can be varied by control of the production conditions to give it a specific composition or content of glucose, maltose or other components, so that the syrups can be 'tailor-made' for specific applications in the food industry.[58,59]

Glucose syrups are classified by reference to their dextrose equivalent (DE) which is the total reducing power expressed as dextrose. The use of these starch-based products as sweeteners instead of sucrose is commercially viable because of the low price of cornstarch. The amounts consumed in the USA and Western Europe are increasing. In 1976, corn sweeteners accounted for 22% of the total annual sugar and sweetener consumption in the USA[60] and 16% in the UK.[3] The situation has now been reached where pressure from sugar producers has caused the EEC to impose harsh economic controls on the starch-sweetener manufacturers, but nevertheless development is proceeding apace, particularly in the further treatment of the basic syrups.

Corn syrups are included in the US Food and Drug Administration

GRAS list as sweetening and thickening agents with the additional properties of retarding crystallisation of sugar in candies, icings and fillings, and preventing the loss of moisture. Because of these properties they are used in a wide variety of manufactured foods.[61]

The nutritional value and physiological effects of glucose syrups have been studied in detail.[49] In summary, they constitute a highly assimilable, osmotically acceptable and palatable source of energy. Fed as the sole dietary carbohydrate, the higher molecular weight glucose-syrup fractions cause a slower blood glucose rise than the simple sugars,[58] and also show important advantages with respect to lipid metabolism in comparison with sucrose (see Section 1 for references).

Little has been published on the effects of glucose syrups on dental health. Once the potential of the syrups for replacing sucrose in certain foods was realised by food manufacturers, a range of foods gradually became available for clinical testing. One of the first was boiled sweets containing no sucrose but either 47·5 or 60·4 % D-glucose, 5·0 % maltose, and either 33·3 % or 46·5 % higher oligosaccharides. After a period of some months storage these experimental sweets were noticeably more hygroscopic than conventional sucrose sweets. Groups of young people were given 450 g of either experimental or conventional sucrose sweets to eat in addition to their normal diet over a three-day period without brushing their teeth. At the end of that time the subjects eating the glucose-syrup sweets had less dental plaque covering their teeth than the sucrose controls. A correlation was found between the tendency to form plaque and dental caries experience. Also the carbohydrate content of the plaque was lower in the glucose-syrup group than the sucrose group.[62]

The next clinical trial tested a low-calorie sweetener made from a high DE glucose syrup consisting almost entirely of glucose. The trial differed from the previous one in that a single group of people was examined after two separate three-day periods using: (1) sucrose, and (2) the new sweetener; however, the results were similar: a reduction in plaque formation on the new sweetener and again a correlation between plaque scores and caries experience.[28] The glucose sweetener contained some saccharin to give it the same sweetness level as sucrose. Under certain circumstances saccharin can suppress the growth of oral micro-organisms,[63,64] and the suggestion was made that the effect observed may have been attributable to the sweeteners saccharin content and not to its carbohydrate composition, but further investigation did not confirm this.

In a third trial, a group of men living in a closed environment (the British

Antarctic Survey) were given a diet in which all the foods were reformulated with a combination of glucose syrups and calcium cyclamate instead of sucrose. This regime was isocaloric with their normal sucrose-containing diet, and the two regimes were compared over periods of 14 weeks. Dental plaque scores fell sharply after 10–12 weeks on the glucose syrup/cyclamate regime, but no such trend was seen on the normal sucrose diet.[65]

These findings, in conjunction with other technological data on the glucose syrups, have been taken to indicate their suitability as ingredients of certain health foods and drinks, e.g. children's fruit syrups and vitamin supplements. Glucose syrup drink supplements may also help to improve physical work performance and response to stress.[66]

Some research has been done in animals on the influence of glucose syrups on dental caries, but one difficulty has been the incorporation of the syrups into the diet for comparison with sucrose and other foods. In one trial in baboons, glucose syrup BPC was spray dried and included in the diet at a level of 75% for comparison with sucrose, fructose and a largely cereal stock diet. Dental plaque was examined at intervals over a six-month period. At the end of this time the glucose-syrup group had less plaque than the fructose group but more than the sucrose group.[67]

Spray-dried glucose syrup was also tested in the diet of rats but did not alter the level of caries significantly compared with sucrose.[68] Using the glucose syrup in its original viscous liquid form incorporated with the powder ingredients of the diet to form a stiff paste did not alter its cariogenicity compared with the spray-dried version. The picture changed, however, when glucose syrup was tested in drinking water and not in solid food.[14] Taken as a 20% solution, glucose syrup gave rise to a level of dental caries 60% lower than sucrose in Osborne–Mendel rats. Dental plaque scores were also significantly reduced. The extent of 'smooth-surface' caries was reduced by 55% compared with sucrose. This is of particular interest in relation to children's dental health; there may be a parallel with a type of rampant caries sometimes seen in young children in whom the labial surface of the enamel is attacked over a wide area when the consumption of sucrose drinks and confectionery is high.

There is very little information available on the behaviour of glucose syrups in dental caries studies *in vitro*. The free glucose they contain will obviously be available for rapid acid production by oral micro-organisms, but, as the chain length of the glucose oligomers increases, breakdown to glucose and its subsequent fermentation will take longer and the likelihood of these oligomer molecules remaining in the mouth at the sites of caries attack long enough to be broken down to cariogenic acids is much reduced.

## 4.2.  High-Fructose Syrups

High-fructose syrups are a development from conventional glucose syrups possessing the advantage of greater sweetness so that they can replace sucrose in certain foods directly.[69] An isomerising enzyme derived from certain micro-organisms is used to convert glucose in corn syrups into fructose. The product typically contains 50 % glucose, 42 % fructose and 8 % oligosaccharides. Such syrups can increase the corn sweetener share of the industrial market because of their equivalent sweetness to sucrose.[37] By 1975 the production of high-fructose corn syrups in the USA exceeded 0·6 million tons per annum.

Specific information on the behaviour of high-fructose syrups in general nutrition and in the mouth is also lacking. All that can be said at the moment is that their behaviour should be predictable from their carbohydrate composition. Fructose metabolism displays certain characteristics distinguishing it from that of other carbohydrates. There are physiological grounds for recommending caution in the wholesale substitution of dietary sucrose, glucose and glucose polymers by fructose at the present time.

## 4.3.  Hydrogenated Glucose Syrups (Lycasins®)

Hydrogenated glucose syrups are believed to hold considerable promise as substitutes for other food carbohydrates. Development was begun in Sweden in 1960 and was continued in the 1970s by Roquette Frères in Lille, France, who now market hydrogenated glucose syrups.

The Lycasins are derived from corn syrup by hydrogenation, resulting in free glucose being converted to sorbitol and the reducing end-units of the oligo- and higher saccharides of glucose being reduced to give sorbitol end-units. As in the case of the glucose syrups, there is a range of products.[70]

The caloric value of the Lycasins is the same as that of starches and sugars. The Lycasins are clear syrups, similar in appearance to glucose syrups, with levels of sweetness approximately 0·75 that of sucrose.

The regulatory position is not clear, but the Lycasins are reported to be already in use in the majority of Western European and Scandinavian countries. In the UK their food-additive status is at present under review.

The range of use of Lycasins is wide. They have been proposed for use in replacing sucrose and glucose syrups in, for example, candy, chewing gum, cough syrup, vitamin syrup, tablets, soft drinks, ice-cream, cornflakes and baked goods. The grade recommended for these applications is Lycasin 80/55, which contains 6–8 % sorbitol, 50–55 % disaccharide, 20–25 %

trisaccharide and 10–20% polysaccharide alcohols.[52] Among the advantages of Lycasins are their anti-crystallising power, non-browning property (even at very high temperatures), and pleasant taste and sweetness characteristics which enable candy to be manufactured with reduced quantities of acid. However, because of their hygroscopicity, individual and special wrappings must be used for the sweets. Some physicochemical and other properties of hydrogenated glucose syrups are described by Kearsley and Birch.[71]

Frostell et al.[72] studied the substitution of sucrose by a Lycasin in candy mainly from the point of view of dental health in children. Occasionally the Lycasin candy caused flatulence and many of the children did not like it. More detailed toxicological studies[52] were carried out by Tacquet in France in 1975 and 1978 and by commercial testing laboratories in the USA in 1969 (it is not clear whether the full results have been published). In the French study, Lycasin drinks were given to 30 men over a three-month period, and it was concluded from a range of tests that biological tolerance was excellent, even in diabetics.

Very little specific information has been found in the literature on the biochemical pathways of metabolism of the Lycasins. Dahlqvist and Telenius[73] worked on some materials very similar to the Lycasins in rats and in vitro, and found that disaccharides containing a molecule of sorbitol linked to a molecule of glucose—either initially present as disaccharides or formed by the enzymic cleavage of oligosaccharides—were hydrolysed very slowly. The glucose bound in this way largely escaped absorption, and the sorbitol, no matter whether free or bound, was absorbed very slowly. The hydrolysis of higher saccharides took place in two steps. The first, brought about by α-amylase, produced di- and oligosaccharides; the second, brought about by intestinal disaccharidases, split dimers containing only glucose to free glucose, which was easily absorbed and utilised by the body.

Research on the dental effects of the Lycasins up to 1977 is summarised by Frostell,[74] but there is a need for more experimental data comparing the Lycasins with the common dietary sugars. Clinical evidence for a benefit from the use of one type of Lycasin rests on the 'Roslagen' study.[72] The Lycasin contained 10·5% sorbitol, 7·5% maltitol and other disaccharide alcohols, 7·0% trimers, 7·0% tetramers, 6·0% pentamers and 62·0% hexamers and higher saccharide alcohols, together with traces of glucose. It was made into various kinds of sweets, including fruit drops, toffee, milk chocolate, lollipops and marshmallow, which were stated to resemble ordinary sweets in taste and appearance, and were the same price. At that time a Lycasin chewing gum could not be made so a sorbitol gum was used instead. Children in the three- to six-year-old age group ate these Lycasin sweets for 18–30 months. A control group ate conventional sucrose sweets. Starting with a total of 225 children, 54 % participated for 18 months and 40 % for 30 months, but more dropped out from the Lycasin than from the control group, and checks showed that substitution of sucrose by Lycasin in the products had been only partial. Measuring the caries increment in the primary dentition, the figures for the control group were consistently greater than those for the Lycasin group, with the disparity between the two widening towards the end of the full 30-month period. At this point the differences between the groups had reached statistical significance although highly significant findings would not be expected from this type of experiment, with only partial substitution of dietary sucrose and relatively small numbers of children. The authors conjectured that a reduction in caries increment of about 25 % was probable.

Sweets made from Lycasins have also been tested in an intra-oral acid measurement technique devised by Mühlemann and his colleagues at the University of Zürich.[53,54] A small pH electrode is set in the position of a tooth surface in a dental prosthetic appliance. It transmits data to a recording device outside the mouth. By observing the fall in pH after a certain food has been in the mouth, Mühlemann classifies it as 'zahn-schonend' (safe for the teeth) or otherwise, and he has obtained official recognition of the test from the Swiss Federal Office of Public Health. Included in the approved lists[75] are various brands of caramels and bonbons based on Lycasin 80/55 with added cyclamate for sweetness.

A Lycasin sample was tested in rodents by Frostell et al.[34] They referred to it as a hydrogenated potato starch but did not specify its composition beyond saying that it contained sorbitol and hydrogenated dextrins (2–10 glucose units) with a sorbitol residue at one end of the chain. Only small numbers of hamsters and rats were used but it was apparent that the inclusion of this product in the diet in place of sucrose was associated with less dental plaque formation and reduced caries development, particularly on smooth surfaces of the teeth as opposed to the molar fissures. Frostell et al. point out that sorbitol can be fermented slowly by certain streptococci associated with the causation of dental caries, and they hesitate to conclude that this particular product is non-cariogenic.

More recent studies just completed at Guy's Hospital, London, show that the cariogenic potential of two different grades of Lycasin is very low in laboratory rats. The levels of caries on the Lycasin regimes were very close to those on corn starch and significantly below the level on a sucrose regime.[76]

These studies *in vivo* have been supplemented by microbiological tests *in vitro* to establish whether Lycasins can be fermented by strains of caries-inducing streptococci and lactobacilli. Acid production from Lycasins, but at a slower rate than from sucrose, was observed by Gehring[77] and Bramstedt and Trautner.[78] Lycasins were fermented by *Streptococcus faecalis*, most lactobacilli, and some strains of *S. sanguis* and *S. mutans*, but not by most other streptococcal strains.[79] The fall in pH of dental plaque on contact with Lycasins may be explained by the existence of other groups of oral micro-organisms which can also attack them.

Studies of acid production from a Lycasin and other carbohydrates were continued by Birkhed and Skude[80] who found that the Lycasin was hydrolysed by salivary amylase with the formation of di- and oligosaccharides, but at a lower rate than starch. Up to 25 % of the cultivable plaque micro-organisms produced extracellular starch-degrading enzymes but the microbial amylase activity was negligible compared with salivary amylase activity.

Finally, incubation with oral micro-organisms produced consistently less acid and less polysaccharide from experimental Lycasin boiled sweets than from conventional sugar-based sweets, and the extent of demineralisation of hydroxyapatite, the mineral matter of dental enamel, was also lower in the presence of the Lycasin sweets.[81]

## 5.  SUGAR ALCOHOLS OR POLYOLS

Chemically, the sugar alcohols are closely related to the sugars, with the reducing group of the sugars replaced by a primary hydroxyl group. Each reducing sugar has a corresponding alcohol, but only three, along with one disaccharide alcohol (maltitol), are practical contenders at present for dietary sucrose substitution.

### 5.1.  Sorbitol

It is logical to follow consideration of the last class of sucrose substitutes, the Lycasins, by sorbitol since the sorbitol content of the Lycasins is a major determinant of their properties.

Sorbitol has the same steric configuration as glucose. It occurs naturally in certain ripe fruit such as cherries, plums, pears and apples. Figures for the consumption of sorbitol are given by Cantor.[82] Sorbitol, frequently blended with mannitol, is used in the USA 'at allowable levels' along with both nutritive and non-nutritive sweeteners to improve flavour, retard crystallisation and to sweeten 'sugarless' confectionery and chewing gum.

Soft, sugar-based candy is said to retain its firmness and chewing properties when 'doctored' with 1–3 % sorbitol, and shelf-life is extended as a result of the inhibition of crystallisation of the sugar. In shredded coconut and other foods, sorbitol not only sweetens but also helps to maintain the moisture content. Small amounts of sorbitol added to low-calorie drinks and foods mask the bitter aftertaste of saccharin and help to provide the 'body' and 'mouthfeel' normally associated with sucrose.[83]

Sorbitol is permitted for use as a food additive by the US Food and Drug Administration, and deemed GRAS for use as a nutrient and dietary supplement; however, it has been designated for additional study.[84]

Some data on the metabolism of sorbitol compared with other sweeteners are given by Bässler.[85] The mean metabolic capacity in man is only about one-third that of glucose and fructose.[86] After relatively slow absorption from the digestive tract, sorbitol is converted to fructose in the liver by sorbitol dehydrogenase, whereafter its further metabolism is the same as for fructose,[87,41] although fructose is not the sole or main intermediate of the metabolic degradation of sorbitol.[88] Because sorbitol is absorbed slowly, blood sugar levels rise only slightly, so that foods sweetened with sorbitol instead of sucrose are said to provide diabetics with a relatively 'safe' source of sweetness and energy; moreover, the metabolism of sorbitol does not involve insulin. However, the total calorie value of foods formulated with sorbitol is as high as that of sucrose-sweetened foods, and medical opinion seems to be swinging away from the liberal use of sorbitol-containing products by diabetics.

Further detailed data on the use of sorbitol are given by Mackay.[89] Sorbitol has 0·5–0·6 times the sweetness of sucrose, but, in spite of its many advantages, its only significant usage in confectionery is for 'sugarless' chewing gum, which could soon account for 25–33 % of the total chewing-gum market. Sorbitol chocolate and boiled sweets can also be made, but the high initial price of sorbitol compared with sucrose, coupled with high processing costs, do not make it an attractive commercial proposition. Furthermore, large amounts of sorbitol (about 50 g in adults), eaten over a short period can lead to osmotic diarrhoea.

As sorbitol is already an accepted food ingredient, there has been little or no incentive for sorbitol manufacturers to promote dental trials, and there is very little in the way of clinical work on record. One trial specifically of sorbitol chewing gum was conducted in eight- to twelve-year-old children in Denmark.[90] After chewing three pieces of the gum a day for two years the caries increment was significantly less (about 10 % less) than in a control group of children, who do not appear to have received sucrose-containing gum for comparison. The authors are cautious about attributing the 10 % reduction in decayed and filled tooth surfaces to the sorbitol gum alone. In addition to 65 % sorbitol, the gum contained 2·4 % calcium phosphate. It was possible that the chewing-gum group may have eaten fewer sucrose-containing sweets than the control group. No changes in gingivitis or dental plaque deposits were observed.

In two other clinical studies 'sugarless' chewing gums were included as controls in monitoring the anti-caries effectiveness of gums containing di-calcium phosphate. After 30 months of chewing five pieces of gum a day, Alabama schoolchildren on the 'sugarless' gum had significantly (16–19 %) lower caries increments than the sucrose gum group. This reduction was on a par with that from the phosphate-containing gum. The detailed composition of the gums is not given but the 'sugarless' gum is believed to be based on sorbitol, possibly in combination with some mannitol.[91]

In a study in British Columbia, in which the composition of all the gums is not given, one of the three experimental groups of children chewed three sticks of sugarless gum a day for two years. The results were inconclusive, however, with no dental benefits shown in the 'sugarless' gum group.[92] Among the limited conclusions that can be drawn from this work, it has been claimed that sugarless gum does not promote tooth decay.

Very little work on the cariogenicity of sorbitol in animals has been reported. Shaw and Griffiths[93] found that 20 % or more of sorbitol replacing part of the 56 % sucrose in a cariogenic diet produced a moderate but consistent reduction in caries in rats, but the rats were liable to diarrhoea if they were given in much more than 20 % dietary sorbitol. The addition of 5 or 10 % sorbitol to high cornstarch diets did not make them more cariogenic, suggesting that sorbitol is less cariogenic than sucrose.[94]

Larje and Larson[95] observed a reduction in smooth-surface caries in rodents when sorbitol instead of sucrose was fed intermittently, but the difference in fissure caries was not clear-cut. At levels of 20–30 % in the diet, sorbitol gave rise to diarrhoea, increased water intake and reduced weight gains.[96] Sorbitol was tested in monkeys by Cornick and Bowen[149] and found to be far less cariogenic than sucrose. It failed to support the growth of *Streptococcus mutans* (considered to be one of the main caries-inducing micro-organisms) in dental plaque, and over a two-year experimental period the capacity of the plaque to produce acid from sorbitol remained unaltered.

Summarising all this animal work, Shaw[94] concluded that sorbitol has 'a generally lower ability to support the carious process in comparison with monosaccharides and disaccharides'. To bring the picture up to date, recent findings from Guy's Hospital confirm that sorbitol is not cariogenic[76,97] and show that the intermittent addition to the diet of a small proportion of sorbitol is sufficient to curb the development of caries in rats receiving a cariogenic sucrose-containing basic diet, compared with the unsupplemented controls.[98]

The usual explanation offered for the lower cariogenicity of sorbitol compared with the sugars is that it cannot be metabolised as easily by the oral microflora to produce the acids that attack the teeth in the caries process. This has led to a large amount of work *in vitro* on the utilisation of sorbitol by cultures of oral micro-organisms.[79,99–102] The findings vary depending on the techniques used and the measurements made, but taking an overview it is clear that oral bacteria that have been allowed no time to adapt, ferment sorbitol less readily than glucose or sucrose, producing far less acid. Adaptation can be contrived, however, by frequent subculturing and exposing the micro-organisms to media containing sorbitol and no sugar. This has provided some grounds for the belief that the formulation of a wide range of foods with sorbitol instead of sucrose would be of limited dental value because micro-organisms in the mouth, if frequently exposed to sorbitol, would sooner or later adapt to ferment it without difficulty.

### 5.2. Mannitol
Mannitol is a hexitol isomeric with sorbitol but with a different steric disposition of the —OH groups. It has approximately 0·4–0·6 times the sweetness of sucrose. It occurs in nature in small quantities and is on the US FDA GRAS list, although designated for additional study.

There is not a great deal in the literature specifically on mannitol. Much of what has been said about sorbitol also applies to mannitol. Sorbitol has received more attention perhaps because of its closer relationship to glucose. As in the case of sorbitol, mannitol is comparatively slowly absorbed from the digestive tract. After absorption it is converted to fructose by a parallel, but separate, metabolic pathway to sorbitol.

The main use of mannitol has been in chewing gum, not as the sole sweetening and bulking agent but blended with sorbitol. In the clinical trials of chewing gum (Section 5.1) the sugarless gums tested in the Alabama and British Columbia studies probably contained mannitol in addition to sorbitol.[91,92]

Some of the animal trials listed under sorbitol also tested mannitol, with similar results and conclusions.[94,95,97] A number of the studies of the metabolism of oral micro-organisms *in vitro* tested mannitol alongside other sucrose substitutes. The results were again closely similar to those with sorbitol but there was some indication that mannitol was fermented even less readily than sorbitol.[102]

From this limited amount of information it is difficult to see how mannitol can possess any significant advantages over sorbitol as a sucrose substitute.

### 5.3. Maltitol

Maltitol is a disaccharide alcohol (4-$\alpha$-D-glucopyranosyl-D-sorbitol) produced by the hydrogenation of maltose and marketed on a small scale in Japan. It is supplied as a syrup containing less than 80% of solids as maltitol, the remainder being polyols of mono-, tri- and oligosaccharides.

The sweetness of maltitol is approximately 0·9 times that of sucrose and it is highly hygroscopic. As it can confer sweetness and body in a similar manner to sucrose, it is said to be suitable for many sorts of confectionery, particularly for good gloss coatings. Such characteristics as stability, inability to undergo the Maillard (browning) reaction, moisture-retention properties and low fermentability by common moulds and bacteria are put forward as points in its favour as a food ingredient.[103]

There is a diversity of opinion on the absorption and metabolism of maltitol. Early Japanese work indicated that very little maltitol was absorbed, but that most of that absorbed was excreted unchanged via the kidney. A high intake in rats produced low growth rates and diarrhoea. Maltitol could therefore be classed as a non-nutritive sweetener.[104] On the other hand it has recently been claimed that maltitol can be metabolised partly to glucose and sorbitol in the stomach, but also to volatile fatty acids by microbes in the large intestine.[105] Studies in other laboratories support the view that mannitol is metabolisable.

The possibility of microbial fermentation has cast some doubt on the earlier claims that maltitol could not be converted by oral micro-organisms to cariogenic acids or polysaccharides and must therefore be harmless to the teeth. On the dental effects of maltitol, all that can be said at present is that further work is needed before any firm statement of advantages as a sucrose substitute can be made.

### 5.4. Isomaltitol

Very little has been published on isomaltitol, a sweet-tasting disaccharide alcohol. As it is one of the two major components of Palatinit, the few studies that have been carried out on its metabolism are discussed under Palatinit (Section 5.6).

## 5.5. Xylitol

Xylitol has been promoted widely over the last few years as a non-cariogenic sweetener, and there is now a very extensive literature on it, ranging from textbooks and proceedings dedicated to xylitol[47,106,107] to collections of papers and monographs on health and sugar substitutes with an emphasis on xylitol.[40,108] This interest in xylitol started with the development of an industrial process in Finland for producing it from wood hemicelluloses by hydrolysis and hydrogenating the liberated xylose, followed by purification. Mass production commenced in Finland in 1974, and research has been stimulated especially within the dental profession.

Xylitol differs from the other monosaccharide alcohols proposed as sucrose substitutes in that it is a pentitol (5C backbone) as opposed to the more common hexitols (e.g. sorbitol and mannitol, 6C backbone). Nevertheless, it occurs in nature in various fruits and vegetables, and is a normal intermediate in carbohydrate metabolism in the human body.

Xylitol is supplied as colourless, non-hygroscopic crystals, and has a sweetness equal to that of sucrose and a pleasant, cool aftertaste. Its calorific value is the same as that of sucrose.[106,109]

The regulatory status of xylitol varies from country to country, and is not clear—within Scandinavia it is permitted in chewing gum and pastilles in Norway but in gum only in Denmark.[110] Xylitol gum also seems to be available for restricted use in certain other countries. It is not mentioned in the US FDA GRAS list but it was stated in 1978 to have FDA and WHO/FAO clearance as an ingredient in dietetic foods.

Although xylitol can be incorporated successfully into a wide variety of manufactured foods in place of sucrose,[44] its main potential use seems to be in confectionery.[111–13] Formulations have been developed to use xylitol in boiled sweets, toffees, caramels, chocolate, compressed tablets, coatings of dragees, and chewing gum, and the results are said to be very good.[114] Among its chief advantages are its microbiological and chemical stability and its behaviour on crystallisation. The xylitol content of some of the lines (percentage by weight) is chocolate 30–43, lozenges 33–58·5, fruit pastilles 40, and candies 43–100. Mäkinen[114] worked out a 'utility index' for foods reformulated with xylitol, taking into account a range of factors including technological advantages, organoleptic properties, compatibility with other foods, dental effects, side effects, shelf-life, price, opportunities for use, and attractiveness, and came to the conclusion that the best products as vehicles for xylitol are pastilles, lozenges, tablets and chewing gum. As it is not expected that the price of xylitol will ever fall sufficiently to be competitive with sucrose, its best prospects will probably lie in special products for the benefit of dental or general health. Some technical problems in the use of xylitol are discussed by Kammerer.[115] Förster[116] cites the following reasons for the current interest in xylitol in nutrition:

1. It is non-toxic;
2. Several species, including man, can handle large doses efficiently, and little of it appears in the urine;
3. The utilisation of xylitol is independent of insulin, suggesting a use as a sweetener for diabetics; and
4. The fact that it is antiketogenic is also relevant to its use for diabetics.

There has been widespread reference in the literature to xylitol in parenteral nutrition,[40] but its use in this way has been queried as a consequence of the results of some Australian work[117] on oxalate crystal deposition in renal, cerebral and vascular tissue.

Although there is an extensive literature on the metabolism of xylitol in the mammalian body, there appears to be very little information about absorption from the gastrointestinal tract. It is likely that special active absorptive mechanisms for xylitol do not exist. Certainly it is known that large amounts eaten at one time can lead to osmotic diarrhoea and disturbances in water balance, as is the case for sorbitol and mannitol, so that not more than 50–70 g daily, spread throughout the day, should be consumed.

Xylitol is metabolised in the liver, first to xylulose and eventually to glucose, lactate or components of the citric acid cycle.[118] There may also be some extrahepatic metabolism. A detailed comparison of the metabolism of xylitol, sorbitol and fructose is given by Förster,[116] and further biochemical data are summarised by Mäkinen.[106] After consideration of the biochemistry of xylitol, Bässler[119] concludes that it is particularly useful for diabetics. Further comments on the tolerance of xylitol by adults and children are made by Förster,[120] pointing out that 10–30 g as a single dose can probably be tolerated by a healthy person without ill effects, and that people on graduated doses can adapt to greater tolerance. When used in confectionery the danger of diarrhoea is not very great, but the intake could be considerably higher if xylitol is used in addition to sweeten soft drinks. The safety of xylitol has also been discussed by Brin and Miller,[121] with the conclusion that it is safe at a daily level of 90 g and beyond, provided it is 'appropriately administered orally with adaptation'.

Some toxicological work on xylitol that gave cause for concern has been summarised in a postscript to a paper by Shaw.[122] Preliminary data showed an increased incidence of calculi in the bladders of mice receiving 10 and 20 % dietary xylitol. In some cases the calculi were associated with other changes including malignant neoplasms, although it was most unlikely that xylitol was the primary carcinogen. No such changes were observed in rats in a parallel study over two years, but there was an increase in adrenal hyperplasia and medullary neoplasia in male rats fed 20 % xylitol. Dogs showed no treatment-related neoplasia. There was also a suspicion that certain of these pathological changes were associated with 20 % sucrose and 20 % sorbitol regimes. Further data are awaited.

The really strong point in favour of xylitol as a dentally beneficial replacement for sucrose is the outcome of major clinical studies conducted in Turku, Finland. The main findings from the Turku studies were presented in a series of 21 papers edited by Scheinin and Mäkinen,[47] but they have been re-reported and reviewed several times.[123-5]

In brief, the Turku two-year feeding study involved the almost complete substitution of sucrose by xylitol in a group of 52 people, which was compared with a control group of 35 on their normal sucrose-containing diet and a group of 38 on a fructose-substituted diet. The average age of the subjects was 27·6 years. In the course of the study 10 people dropped out, but in only one case was this due to osmotic diarrhoea from xylitol. All the groups started off with about the same level of dental caries, and the experimental groups used foods reformulated with xylitol or fructose instead of sucrose.[44]

Based on clinical and radiographic assessment at the end of the first year, a highly significant reduction in caries (85 %) was found in the xylitol group compared with the sucrose control. (The corresponding reduction with fructose was much smaller—see Section 3.2.) After two years the mean increment in the number of decayed, missing and filled tooth surfaces as a consequence of caries was zero in the xylitol group, compared with 7·2 in the sucrose control group. Many other confirmatory data are given in the reports of the study, and the graphs of the results of the dental examinations conducted every few months provide convincing evidence of deterioration in dental health in the sucrose group while the xylitol group showed little or no increase in caries. In fact between the 20th and the 24th month the number of decay lesions actually appeared to decline on the xylitol regime. This led to xylitol being hailed in some quarters as a possible caries-reversing agent but, seen in perspective, this trend makes only a small contribution to the major finding that the xylitol regime did not promote caries.

A second study was conducted at Turku on the effects of xylitol versus ordinary sucrose-containing chewing gum over a one-year period with 102 subjects of mean age 22·2 years. They ate their normal diet but chewed, in addition, an average of four to five pieces of gum spaced throughout the day. The pieces of gum weighed 3 g, of which 1·5 g was either xylitol or sucrose. At the end of the year the number of decayed, missing and filled tooth surfaces had risen in the sucrose gum group but had fallen in the xylitol group by a highly significant margin. The carious surface increment was 3·76 in the sucrose group and 0·33 in the xylitol group. Again there appeared to be some reversals of caries in the final few months but the findings are not conclusive on this point. In summary, Scheinin et al.[126] concluded that the findings clearly indicated a therapeutic, caries-inhibitory effect of xylitol. Swango[124] has pointed out, however, that it does not seem appropriate to attribute 'negative increments' to a therapeutic effect in the absence of replicate findings or other strongly corroborative evidence, especially since xylitol in gum was compared only with sucrose gum which is known to be cariogenic. Because a control non-cariogenic gum was not included in the study, it cannot be determined whether the differences recorded were attributable to increased caries activity in the sucrose group, caries inhibition in the xylitol group, or a combination of both. Many subsidiary studies were done to monitor any associated biochemical, metabolic, salivary or bacterial changes in the Turku trials.[47]

No further reports of large-scale trials of xylitol products in man have appeared. In the early 1980s the US National Institute of Dental Research was said to be funding a trial of xylitol chewing gum in children but the project is now in abeyance. Another trial of xylitol confectionery is in hand in the Far East.

Some short-term studies on dental plaque development after chewing xylitol gum have been reported. Compared with sucrose gum, Mouton et al.[127] found that xylitol gum was associated with less plaque, but other unpublished work reviewed by Scheinin[125] failed to substantiate this. In groups of men given six pieces of gum to chew per day over a three-day period, there was significantly less plaque on the teeth at the end of the xylitol than the sucrose gum period. The significance of this lies in the correlation which was found between the individuals' tendency to form plaque and their caries experience.[128]

The encouraging results from Turku stimulated a number of laboratory animal trials, reviewed by Scheinin[125] and Swango.[124] Compared with sucrose, xylitol was not cariogenic in any of the trials, and two of them even showed the surprising finding that partial substitution of dietary sucrose by xylitol could reduce the cariogenicity of sucrose.[129,130] Other reports did not confirm this anti-caries activity.[96,124] To complete the picture, when xylitol was incorporated at three different levels in a cariogenic sucrose-containing diet, 20 % xylitol in place of sucrose reduced the level of caries in rats by the same margin as 20 % wheatstarch, indicating that both are non-cariogenic but that xylitol did not exert an active caries-inhibiting effect. Replacing 10 % of the sucrose in the cariogenic diet by xylitol produced virtually the same level of caries as the basic cariogenic diet, but xylitol added to the basic diet at 2 % was associated with an increase in caries, possibly as a result of greater food consumption or frequency of eating. Further trials confirmed that xylitol neither promoted nor reversed dental caries, and investigated adaptation of the rats to increasing levels of dietary xylitol.[97]

These trials in animals were done under a variety of different conditions and dietary regimes. Among subsidiary findings, are that the animals may eat a xylitol (or other polyol) diet less readily than the corresponding sucrose diet, and a dietary content of approximately 20 % or more given to young rats without adaptation can lead to gastrointestinal disturbance, diarrhoea, increased water consumption, weight loss and debility. On the whole, the existence of a significant anti-cariogenic effect of xylitol in experimental animals requires more evidence but there are good grounds for describing it as inert with respect to dental caries.

Finally, turning to studies *in vitro*, there has been a considerable amount of work done since 1976 which mostly compares the bacterial metabolism of xylitol with other polyols and sucrose substitutes.[100,102,106,131,132] The overall impression from these and many other investigations is that xylitol is less readily metabolised by oral micro-organisms than the hexitols sorbitol and mannitol. Early work indicated that xylitol is fermented very little or not at all, but claims that it is completely non-fermentable by oral micro-organisms should be regarded with caution. From among a range of streptococci and lactobacilli, Edwardsson *et al.*[79] found two strains that could ferment xylitol without adaptation. In general, however, it seems to be easier for bacteria to adapt to metabolise the hexitols than xylitol.

The addition of xylitol to culture media containing glucose or sucrose slowed the growth of certain bacteria but had little influence on the final pH.[100] Gehring[133] cited a few strains of bacteria that can catabolise xylitol and proceeded to link microbiological studies with levels of caries in gnotobiotic rats fed on specified diets. Although he inoculated some of the rats with a xylitol-metabolising streptococcus, this did not produce noteworthy caries scores when the rats were fed on a xylitol-containing diet.

## 5.6. Palatinit

As with several of the other sucrose substitutes, it has been difficult to collect much information about Palatinit, but some general data about it are given by Siebert and Grupp.[134] Their description of the production method of the Süddeutsche Zucker–AG is not absolutely clear, but starting from sucrose it seems to involve the microbiological conversion of sucrose to isomaltulose and reduction of the fructose moiety to give a mixture of 6-glucosidyl-sorbitol and mannitol.[135] The product is a mixture of 6-$o$-$\alpha$-D-glucopyranosyl-sorbitol (isomaltitol) and 6-$o$-$\alpha$-D-glucopyranosyl-mannitol. Some of its chemical properties are given by Gau et al.[136] and further details of the production method and technological properties are given by Schiweck.[137]

The sweetening power of Palatinit is 0·45 that of sucrose. The taste profile is 'pure sweet' without the cooling effect of xylitol and sorbitol. In confectionery manufacture it is important to note that the melting points of the components of Palatinit are of the same order as that of sucrose. There is no Maillard reaction. After outlining solubility behaviour, viscosity and hygroscopicity (which is very low), Schiweck[137] lists chocolate slabs, drops, bars, marzipan products, and chewing gum among the items in which Palatinit has been tested on a 'semi-technical' scale.

The high stability of Palatinit to hydrolysis is said to be an important characteristic in relation to its cariogenic potential and nutritional physiology. Palatinit can be catabolised by certain enzymes in the human gut but digestion in the small intestine is incomplete and, as with other polyol-containing sweeteners, diarrhoea can ensue if doses as large as 100 g are taken by adults. The final faecal excretion of Palatinit is low, however, as a result of its fermentation by bacteria in the upper colon. The slow rate of assimilation means that it does not produce a significant rise in blood sugar or plasma insulin. These and other features of the metabolism of Palatinit are discussed by Grupp and Siebert.[138] Weight for weight in the diet, Palatinit produces less energy than the common dietary carbo-hydrates sucrose, and starch.

Since one of the two constituents of Palatinit is isomaltitol, studies of the metabolic fate of isomaltitol are relevant here. Partial absorption from the intestine was followed by partial utilisation for energy production in growing rats but the extent of utilisation declined with increasing doses. When isomaltitol was fed for several weeks the caecum enlarged, but it was nevertheless found to be well tolerated by the rat.[139] Basic metabolic data in man demonstrating a low rate of enzymic hydrolysis and slow utilisation were reported by Siebert *et al.*[140]

Information is gradually becoming available on the effect of Palatinit on dental caries in animals. Karle and Gehring[141,142] noted that Palatinit was less cariogenic in rats than sucrose and lactose, and that strains of *Streptococcus mutans* did not produce much acid or extracellular polysaccharide from it. From these and later microbiological and rat experiments, Karle and Gehring concluded that Palatinit seems to be suitable as a sugar substitute for caries prevention.

### 5.7. Lactitol

Lactitol (4-*o*-β-D-galactopyranosyl-D-glucitol) is a disaccharide alcohol produced by the reduction of lactose. Little published information has been found. Aminoff[143] mentions lactitol, in a section on maltitol, as an analogous hydrogenation product with the same non-caloric claim as maltitol. Its low sweetness is said to make it less attractive as a sweetener than maltitol (Section 5.3). Beereboom[56] cites one piece of work by Malsuo[144] using oral micro-organisms, which indirectly suggests that lactitol might be non-cariogenic.

## 6.   OTHER PRODUCTS

### 6.1.  Coupling Sugar

Coupling sugar is a novel material which is something of a hybrid between a sugar and a starch derivative; it is made in Japan by reacting sucrose with starch in the presence of an enzyme called cyclodextrin glycosyl transferase. This results in the removal of one or more glucose units from the starch to link to the glucose moiety of the sucrose molecule. The two principal constituents of coupling sugar are glucosylsucrose (G2F), which has one extra glucose unit attached to the sucrose molecule, and maltosylsucrose (G3F, two extra glucose units attached to sucrose). It is said to be cheap to produce and to have a taste like cane sugar with a sweetness value of 0·6 relative to sucrose.

Following satisfactory safety testing, coupling sugar is distributed mainly as a syrup with a moisture content of 25% for use as a low-cariogenic sweetener in confectionery, other foods and beverages, but only within Japan. It is said to be metabolisable in the same way as other carbohydrates, following the same enzymic hydrolysis, absorption and utilisation patterns.

Foods containing coupling sugar produced less dental plaque than their sucrose counterparts[145] and also less acid.[146] Indications in rats were that coupling sugar at 5% in the diet was less cariogenic than sucrose.[147] *Streptococcus mutans*, which is among the causative organisms of dental caries, produces less acid and less polysaccharide (glucan) from coupling sugar than from sucrose.[148]

## 7. CONCLUSIONS

It is clear from the consideration of the individual sweetening agents that each has certain points in favour and certain shortcomings. In some instances a great deal of additional information is required before wide-scale use can be contemplated, and it is known that research is in hand to answer questions on toxicity, nutritional value and dental effects; however, in other cases the exact situation is unknown and there is a scarcity of published data.

From the limited amount of information available it can be seen that the sugars glucose, fructose and maltose but not L-sorbose (Section 3) have the advantage that they are all well-known, accepted ingredients of foods, and that there are no queries about their metabolism. The evidence suggests that they are somewhat less cariogenic than sucrose but it is not easy to decide if this difference is great enough to warrant replacing sucrose in items of confectionery by any of the three sugars. Certainly such products would not qualify for a claim of 'non-cariogenic' or 'cannot harm the teeth'.

The starch hydrolysates (Section 4) offer better prospects. The physiological effects and metabolism of the glucose syrups have been studied in detail and show no adverse features. They are already in use to a considerable extent in everyday foods including various types of confectionery, frequently blended with sucrose. Such work as has been done suggests that they are less harmful to the teeth than sucrose. In price they can be competitive with sucrose. Thus it seems likely that the use of glucose syrups will continue to grow, particularly because of their advantages for the food manufacturer. Too little is known of the physiological and dental effects of the high-fructose syrups for any comment to be made on them at present.

A good deal of information is now available on the Lycasins. Nutritional and physiological data are generally favourable. Dental data have been accumulating over the last few years but further clinical and laboratory animal trials would be helpful. To interject one note of caution, because of their content of sorbitol and reduced saccharide end-units, unrestricted intake of Lycasins above a certain level can give rise to digestive disturbances, as in the case of polyols.

Among the polyols, sorbitol and xylitol have received the most attention. The metabolism of sorbitol has been studied thoroughly as a result of its usefulness for diabetics. Dental data are limited but there are strong indications that sorbitol is less cariogenic than sucrose. Its main use in confectionery so far seems to be in 'sugarless' chewing gum. Mannitol is also used in this type of gum but it does not appear to have any advantages over sorbitol as a sucrose substitute.

Maltitol, Palatinit and lactitol have not yet reached the stage where they can be put forward as realistic alternatives to sucrose in confectionery. In contrast, xylitol has been studied in depth and a wealth of data are available in the literature. Xylitol has shown no obvious nutritional or physiological drawbacks provided it is consumed below the level that causes gastrointestinal upset. The dental findings are impressive as they stand but will be much strengthened if supporting clinical data from outside Finland and additional laboratory animal data become available. The caries-inhibitory action claimed at one time now seems to be less likely.

Many items of our daily diet, by virtue of their sucrose content and physical characteristics, contribute to dental decay, and of course the situation is exacerbated by unbalanced eating habits and overconsumption. The final choice of a non-cariogenic or less cariogenic sweetener/bulking agent for replacing sucrose rests with the food manufacturer, whose decision obviously depends on many other factors besides dental considerations. It is apparent that no one agent will fulfil all the requirements for the entire range of confectionery products. The characteristis of each type of product must be analysed to determine which agent can be substituted for sucrose most successfully.

## REFERENCES

1. CADBURY LTD (1979). *Confectionery Market Review.*
2. COCOA, CHOCOLATE AND CONFECTIONERY ALLIANCE (1981). *Annual Report 1980–81.*
3. MINISTRY OF AGRICULTURE, FISHERIES AND FOOD (1977). *Food Facts No. 7,* MAFF, London, 2.
4. MASTERS, D. H. and LEWIS, H. (1975). *J. Amer. Soc. for Preventitive Dentistry,* **5**, 23.
5. CLEAVE, T. L. and CAMPBELL, G. D. (1966). *Diabetes, Coronary Thrombosis and the Saccharine Disease,* John Wright & Sons Ltd., Bristol.

6. YUDKIN, J. (1972). *Pure, White and Deadly*, Davis-Poynter, London.
7. AGRICULTURAL RESEARCH COUNCIL/MEDICAL RESEARCH COUNCIL (1974). *Food and Nutrition Research*. HMSO, London.
8. GARROW, J. S. (1978). *Energy Balance and Obesity in Man*, Elsevier, Amsterdam.
9. DEPARTMENT OF HEALTH AND SOCIAL SECURITY (1976). *Research on Obesity*, HMSO, London.
10. MACDONALD, I. (1962). *J. Physiol.*, **162**, 334.
11. ALLEN, R. J. L. and LEAHY, J. S. (1966). *Brit. J. Nutr.*, **20**, 339.
12. BROOK, M. and NOEL, P. (1969). *Nature, London*, **222**, 562.
13. LANDES, D. R. and MILLER, J. (1976). *Cereal Chem.*, **53**, 678.
14. GRENBY, T. H. and LEER, C. J. (1974). *Caries Research*, **8**, 368.
15. MACDONALD, I., GRENBY, T. H., FISHER, M. A. and WILLIAMS, C. (1981). *J. Nutrition*, **111**, 1543.
16. FINN, S. B. and GLASS, R. L. (1975). *World Review of Nutrition and Dietetics*, **22**, 304.
17. NEWBRUN, E. (1967). *Odont. Rev*, **18**, 373.
18. GRENBY, T. H. (1980). *Nutrition and Food Science*, **63**, 9; **64**, 19.
19. GRAY, P. G., TODD, J. E., SLACK, G. L. and BULMAN, J. S. (1970). *Adult Dental Health in England and Wales in 1968*, HMSO, London.
20. TODD, J. E. and WHITWORTH, A. (1974). *Adult Dental Health in Scotland 1972*, HMSO, London.
21. SUTCLIFFE, P. (1972). *J. Periodontal Research*, **7**, 52.
22. BROOK, M. (1973). In: *Sugar: Chemical, Biological and Nutritional Aspects of Sucrose*. J. Yudkin, J. Edelman and L. Hough (Eds.), Butterworth, London, 32–46.
23. HOSKINS, W. A. (1978). In: *Proceedings—Sweeteners and Dental Caries. Special Supplement. Feeding, Weight and Obesity Abstracts*, J. H. Shaw and G. G. Roussos (Eds.), Information Retrieval Ltd, London, 371–80.
24. CAKEBREAD, S. (1975). *Sugar and Chocolate Confectionery*, Oxford University Press, London.
25. MOSKOWITZ, H. R. (1974). In: *Sugars in Nutrition*, H. L. Sipple and K. W. McNutt (Eds.), Academic Press, New York, 37–64.
26. CANTOR, S. M. (1975). In: *Sweeteners. Issues and Uncertainties*, Natl. Acad. of Sciences, Washington, DC, 19–29.
27. HORECKER, B. L. (1976). In: *Monosaccharides and Polyalcohols in Nutrition, Therapy and Dietetics*, G. Ritzel and G. Brubacher (Eds.), Huber, Berne, 1–21.
28. GRENBY, T. H. (1975). *Brit. Dent. J.*, **139**, 129.
29. COLE, J. A. (1977). *Biochem. Soc. Transactions*, **5**(4), 1232.
30. SHAFER, W. G. (1949). *Science, N.Y.*, **110**, 143.
31. GUSTAFSON, G., STELLING, M., ABRAMSON, E. and BRUNIUS, E. (1955). *Odont. Tidskr.*, **63**, 506.
32. GRENBY, T. H. (1963). *Archs. Oral Biol.*, **8**, 27.
33. SHAW, J. H., KRUMINS, I. and GIBBONS, R. J. (1967). *Archs. Oral Biol.*, **12**, 755.
34. FROSTELL, G., KEYES, P. H. and LARSON, R. H. (1967). *J. Nutrition*, **93**, 65.
35. GREEN, R. M. and HARTLES, R. L. (1969). *Archs. Oral Biol.*, **14**, 235.
36. GRENBY, T. H. and HUTCHINSON, J. B. (1969). *Archs. Oral Biol.*, **14**, 373.

84    T. H. GRENBY

37. SEIDMAN, M. (1977). In: *Developments in Food Carbohydrate—I*, G. G. Birch and R. S. Shallenberger (Eds.), Applied Science Publishers Ltd, London, 19–42.
38. OSBERGER, T. F. and LINN, H. R. (1978). In: *Low Calorie and Special Dietary Foods*, B. K. Dwivedi (Ed.), CRC Press, Florida, 115–23.
39. PAWAN, G. L. S. (1973). In: *Molecular Structure and Function of Food Carbohydrate*, G. G. Birch and L. F. Green (Eds.), Applied Science Publishers Ltd, London, 65–80.
40. RITZEL, G. and BRUBACHER, G. (Eds.) (1976). *Monosaccharides and Polyalcohols in Nutrition, Therapy and Dietetics. International J. Vitamin and Nutrition Res.*, No. 15, Huber, Berne.
41. VAN DEN BERGHE, G. (1978). In: *Health and Sugar Substitutes*, Proc. ERGOB Conference, Geneva, B. Guggenheim (Ed.), Karger, Basel, 92–7.
42. PALM, J. D. (1975). In: *Physiological Effects of Food Carbohydrates*, ACS Symposium Series 15, A. Jeanes and J. Hodge (Eds.), Amer. Chem. Soc., Washington D.C., 54–72.
43. SCHEININ, A. (1978). In: *Health and Sugar Substitutes*, Proc. ERGOB Conference, Geneva, B. Guggenheim (Ed.), Karger, Basel, 241–6.
44. MÄKINEN, K. K. and SCHEININ, A. (1975). *Acta Odont. Scand.*, **33**, Supplement 70, 105.
45. SCHEININ, A., MÄKINEN, K. K. and YLITALO, K. (1975). *Acta. Odont. Scand.*, **33**, Supplement 70, 5.
46. SCHEININ, A., MÄKINEN, K. K. and YLITALO, K. (1975). *Acta. Odont. Scand.*, **33**, Supplement 70, 67.
47. SCHEININ, A. and MÄKINEN, K. K. (Eds.) (1975). *Acta. Odont. Scand.*, **33**, Supplement 70, 1.
48. PAGE, L. and FRIEND, B. (1974). In: *Sugars in Nutrition*, H. L. Sipple and K. W. McNutt (Eds.), Academic Press, New York, 93–107.
49. MACDONALD, I. (1974). In: *Sugars in Nutrition*, H. L. Sipple and K. W. McNutt (Eds.), Academic Press, New York, 303–13.
50. GRENBY, T. H. and BULL, J. M. (1977). In: *Developments in Food Carbohydrate—I*, G. G. Birch and R. S. Shallenberger (Eds.), Applied Science Publishers Ltd, London, 169–84.
51. GRENBY, T. H. and PATERSON, F. M. (1972). *Brit. J. Nutr.*, **27**, 195.
52. ZIMMERMAN, M. (1978). In: *Health and Sugar Substitutes*, Proc. ERGOB Conference, Geneva, B. Guggenheim (Ed.), Karger, Basel, 145–52.
53. GRAF, H. and MÜHLEMANN, H. R. (1966). *Helv. Odont. Acta*, **10**, 94.
54. BRUNNER, K., GABATHULER, H. and SCHNEIDER, P. H. (1974). *Helv. Odont. Acta*, **18**, 83.
55. SIEBERT, G., ROMEN, W., SCHNELL-DOMPERT, E. and HANNOVER, R. (1980). *Infusions Therapie und Klinische Ernährung*, **7**, 3.
56. BEEREBOOM, J. J. (1978). In: *Low Calorie and Special Dietary Foods*, B. K. Dwivedi (Ed.), CRC Press Inc., Florida, 39–49.
57. HUCHETTE, M. and LEROY, P. (1974). German Patent 2, 308, 163; C.A. 81, 76699.
58. BIRCH, G. G. (1977). In: *Developments in Food Carbohydrate—I*, G. G. Birch and R. S. Shallenberger (Eds.), Applied Science Publishers Ltd, London, 1–17.

59. MADSEN, G. V. and NORMAN, B. E. (1973). In: *Molecular Structure and Function of Food Carbohydrate*, G. G. Birch and L. F. Green (Eds.), Applied Science Publishers Ltd, London, 50–64.
60. WALTER, B. J. (1978). In: *Proceedings—Sweeteners and Dental Caries. Special Supplement, Feeding, Weight and Obesity Abstracts*, J. H. Shaw and G. G. Roussos (Eds.), Information Retrieval Ltd, London, 93–109.
61. BIRCH, G. G., GREEN, L. F. and COULSON, C. B. (1970). *Glucose Syrups and Related Carbohydrates*, Elsevier, Amsterdam.
62. GRENBY, T. H., POWELL, J. M. and GLEESON, M. J. (1974). *Archs. Oral Biol.*, 19, 217.
63. LINKE, H. A. B. and CHANG, C. A. (1976). *Z. Natursforsch.*, 31c, 245.
64. GRENBY, T. H. and BULL, J. M. (1979). Abstract No. 16, ORCA Congress, Turku, 1978, *Caries Research*, 13, 89.
65. FRY, A. J. and GRENBY, T. H. (1972). *Archs. Oral Biol.*, 17, 873.
66. BROOKE, J. D. (1973). In: *Molecular Structure of Food Carbohydrate*, G. G. Birch and L. F. Green (Eds.), Applied Science Publishers Ltd, London, 235–61.
67. GRENBY, T. H. (1971). *Archs. Oral Biol.*, 16, 631.
68. GRENBY, T. H. (1972). *Caries Research*, 6, 52.
69. MITCHELL, E. L. (1974). In: *Sugars in Nutrition*, H. L. Sipple and K. W. McNutt (Eds.), Academic Press, New York, 127–33.
70. LEROY, P. (1978). In: *Health and Sugar Substitutes*, Proc. ERGOB Conference Geneva, B. Guggenheim (Ed.), Karger, Basel, 114–19.
71. KEARSLEY, M. W. and BIRCH, G. G. (1979). In: *Sugar Science and Technology*, G. G. Birch and K. J. Parker (Eds.), Applied Science Publishers Ltd, London, 287–306.
72. FROSTELL, G. *et al.* (1974). *Acta Odont. Scand.*, 32, 235.
73. DAHLQVIST, A. and TELENIUS, U. (1965). *Acta Physiol. Scand.*, 63, 156.
74. FROSTELL, G. (1977). *Dtsch. Zahnärztl. Z.*, 32, Suppl. 1, S71.
75. Eidgenössisches Gesundheitsamt, Bern. (1978). *EGA—Bewilligte Bonbons auf dem Schweizer Markt/Stand*, Januar.
76. GRENBY, T. H. (1982). Abstract no. 191, I.A.D.R. British Division Meeting, April, *J. Dent. Res.*, 61, 557.
77. GEHRING, F. (1971). *Dtsch. Zahnärztl. Z.*, 26, 1162.
78. BRAMSTEDT, F. and TRAUTNER, K. (1971). *Dtsch. Zahnärztl. Z.*, 26, 1135.
79. EDWARDSSON, S., BIRKHED, D. and MEJÀRE, B. (1977). *Acta Odont. Scand.*, 35, 257.
80. BIRKHED, D. and SKUDE, G. (1978). *Scand. J. Dent. Res.*, 86, 248.
81. GRENBY, T. H. and SALDANHA, M. G. (1981). Abstract no. 211, I.A.D.R. British Division Meeting, April, *J. Dent Res.*, 60(B), 1192.
82. CANTOR, S. M. (1978). In: *Proceedings—Sweeteners and Dental Caries. Special Supplement, Feeding, Weight and Obesity Abstracts*, J. H. Shaw and G. G. Roussos (Eds.), Information Retrieval Ltd, London, 111–28.
83. JACOBSON, M. F. (1976). *Eater's Digest. The Consumer's Factbook of Food Additives*, Doubleday and Co. Inc., New York.
84. SULLIVAN, G. (1976). *Additives in your Food*, Cornerstone, New York.
85. BÄSSLER, K. H. (1976). In: *Monosaccharides and Polyalcohols in Nutrition, Therapy and Dietetics*, G. Ritzel and G. Brubacher (Eds.), Huber, Berne, 22–30.

86. MEHNERT, H., DIETZE, G. and HASLBECK, M. (1975). *Nutr. Metabol.*, **18** (Supplement 1), 171.

87. TOUSTER, O. (1975). In: *Physiological Effects of Food Carbohydrates*, A. Jeanes and J. Hodge (Eds.), Amer. Chem. Soc., Washington D.C., 123–134.

88. SCHNELL-DOMPERT, E. and SIEBERT, G. (1980). *Hoppe-Seyler's Z. Physiol. Chem.*, **361**, 1069.

89. MACKAY, D. A. M. (1978). In: *Health and Sugar Substitutes*, Proc. ERGOB Conference Geneva, B. Guggenheim (Ed.), Karger, Basel, 124–9.

90. MØLLER, I. J. and POULSEN, S. (1973). *Community Dent. Oral Epidemiol.*, **1**, 58.

91. FINN, S. B. and JAMISON, H. C. (1967). *J. Amer. Dent. Assoc.*, **74**, 987.

92. RICHARDSON, A. S., HOLE, L. W., McCOMBIE, F. and KOLTHAMMER, J. (1972). *J. Canadian Dental Assoc.*, **38**, 213.

93. SHAW, J. H. and GRIFFITHS, D. (1960). *J. Dent. Res.*, **39**, 377.

94. SHAW, J. H. (1976). *J. Dent. Res.*, **55**, 376.

95. LARJE, O. and LARSON, R. H. (1970). *Archs. Oral Biol.*, **15**, 805.

96. MÜHLEMANN, H. R., REGOLATI, B. and MARTHALER, T. M. (1970). *Helv. Odont. Acta*, **14**, 48.

97. GRENBY, T. H. and COLLEY, J. (1983). *Arch. Oral Biol.* (In press).

98. GRENBY, T. H. (1983). Abstract 81, Proceedings of ORCA Congress, June and *Caries Res.*, **17**, 185.

99. BROWN, A. T. (1977). In: *Proceedings—Workshop on Cariogenicity of Food, Beverages, Confections and Chewing Gum*, Amer. Dent. Assoc. Health Foundation, Chicago, 114–18.

100. HAVENAAR, R., HUIS IN'T VELD, J. H. J., BACKER DIRKS, O. and DE STOPPELAAR, J. D. (1978). In: *Health and Sugar Substitutes*, Proc. ERGOB Conference, Geneva, B. Guggenheim (Ed.), Karger, Basel, 192–8.

101. GÜLZOW, H. J. (1976). In: *Monosaccharides and Polyalcohols in Nutrition, Therapy and Dietetics*, G. Ritzel and G. Brubacher (Eds.), Huber, Berne, 348–57.

102. MÄKINEN, K. K. (1976). In: *Microbial Aspects of Dental Caries*, H. M. Stiles, W. J. Loesche and T. C. O'Brien (Eds.), Information Retrieval Ltd, London, 521–38.

103. NAITO, F. (1971). *New Food Industry*, **13**, 1.

104. OKU, T., INDUE, Y. and HOSOYA, N. (1971). *Eiyo To Shokuryo*, **24**, 399.

105. RENNHARD, H. and BIANCHINE, J. R. (1976). *J. Agric. Food Chem.*, **24**, 387.

106. MÄKINEN, K. K. (1978). Biochemical Principles of the Use of Xylitol in Medicine and Nutrition, with Special Consideration of Dental Aspects, *Experientia*, Supplement 30, Birkhäuser Verlag, Basel.

107. COUNSELL, J. N. (Ed.) (1978). *Xylitol*, Applied Science Publishers Ltd, London.

108. SHAW, J. H. and ROUSSOS, G. G. (Eds.) (1978). *Proceedings—Sweeteners and Dental Caries. Special Supplement, Feeding, Weight and Obesity Abstracts*, Information Retrieval Ltd, London.

109. AMINOFF, C., VANNINEN, E. and DOTY, T. E. (1978). In: *Xylitol*, J. N. Counsell (Ed.), Applied Science Publishers Ltd, London, 1–9.

110. POULSEN, E. (1978). In: *Health and Sugar Substitutes*, Proc. ERGOB Conference Geneva, B. Guggenheim (Ed.), Karger, Basel, 277–80.

111. VOIROL, F. (1978). In: *Xylitol*, J. N. Counsell (Ed.), Applied Science Publishers Ltd, London, 11–20.
112. VOIROL, F. (1978). In: *Health and Sugar Substitutes*, Proc. ERGOB Conference, Geneva, B. Guggenheim (Ed.), Karger, Basel, 130–7.
113. VOIROL, F. A. (1979). In: *Sugar Science and Technology*, G. G. Birch and K. J. Parker (Eds.), Applied Science Publishers Ltd, London, 325–44.
114. MÄKINEN, K. K. (1977). In: Proceedings of Workshop on Cariogenicity of Food, Beverages, Confections and Chewing Gum, Amer. Dent. Assoc. Health Foundation, Chicago, 99–113.
115. KAMMERER, F. X. (1977). *Dtsch. Zahnärztl. Z.*, **32**, Suppl. 1, S19.
116. FÖRSTER, H. (1974). In: *Sugars in Nutrition*, H. L. Sipple and K. W. McNutt (Eds.), Academic Press, New York, 259–80.
117. THOMAS, D. W., EDWARDS, J. B., GILLIGAN, J. E., LAWRENCE, J. R. and EDWARDS, R. G. (1972). *Med. J. Aust.*, **1**, 1238.
118. FROESCH, E. R. and JAKOB, A. (1974). In: *Sugars in Nutrition*, H. L. Sipple, and K. W. McNutt (Eds.), Academic Press, New York, 241–58.
119. BÄSSLER, K. H. (1978). In: *Xylitol*, J. N. Counsell (Ed.), Applied Science Publishers Ltd, London, 35–41.
120. FÖRSTER, H. (1978). In: *Xylitol*, J. N. Counsell (Ed.), Applied Science Publishers Ltd, London, 43–66.
121. BRIN M. and MILLER, O. N. (1974). In: *Sugars in Nutrition*, H. L. Sipple and K. W. McNutt (Eds.), Academic Press, New York, 591–606.
122. SHAW, J. H. (1987). In: *Proceedings—Sweeteners and Dental Caries. Special Supplement, Feeding, Weight and Obesity Abstracts*, J. H. Shaw and G. G. Roussos (Eds.), Information Retrieval Ltd, London, 157–76.
123. MÄKINEN, K. K. (1978). In: *Proceedings—Sweeteners and Dental Caries. Special Supplement, Feeding, Weight and Obesity Abstracts*, J. H. Shaw and G. G. Roussos (Eds.), Information Retrieval Ltd, London, 193–220.
124. SWANGO, P. A. (1978). In: *Proceedings—Sweeteners and Dental Caries. Special Supplement, Feeding, Weight and Obesity Abstracts*, J. H. Shaw and G. G. Roussos (Eds.), Information Retrieval Ltd, London, 225–34.
125. SCHEININ, A. (1978). In: *Xylitol*, J. N. Counsell (Ed.), Applied Science Publishers Ltd, London, 139–61.
126. SCHEININ, A., MÄKINEN, K. K., TAMMISALO, E. and REKOLA, M. (1975). *Acta Odont. Scand.*, **33**, Suppl. 70, 307.
127. MOUTON, C., SCHEININ, A. and MÄKINEN, K. K. (1975). *Acta Odont. Scand.*, **33**, 251.
128. GRENBY, T. H., BASHAARAT, A. H. and GEY, K. F. (1982). *Brit. Dent. J.*, **152**, 339.
129. MOLL, R. and BÜTTNER, W. (1978). *Caries Res.*, **12**, 119.
130. LEACH, S. A. and GREEN, R. M. (1980). *Caries Res.*, **14**, 16.
131. CARLSSON, J. (1978). In: *Health and Sugar Substitutes*, Proc. ERGOB Conference on Sugar Substitutes, Geneva, B. Guggenheim (Ed.), Karger, Basel, 205–10.
132. BIRKHED, D. and EDWARDSSON, S. (1978). In: *Health and Sugar Substitutes*, Proc. ERGOB Conference on Sugar Substitutes, Geneva, B. Guggenheim (Ed.), Karger, Basel, 211–17.

133. GEHRING, F. (1978). In: *Health and Sugar Substitutes*, Proc. ERGOB Conference Geneva, B. Guggenheim (Ed.), Karger, Basel, 229–34.
134. SIEBERT, G. and GRUPP, U. (1978). In: *Health and Sugar Substitutes*, Proc. ERGOB Conference, Geneva, B. Guggenheim (Ed.), Karger, Basel, 109–13.
135. SCHIWECK, H., STEINLE, G. and HABERL, L. (1975). US Patent, 3,865,957 Sueddeutsche Zucker-AG, Mannheim, West Germany.
136. GAU, W. *et al.* (1979). *Z. Lebensm. Unters. Forsch.*, **168**, 125.
137. SCHIWECK, H. (1978). In: *Health and Sugar Substitutes*, Proc. ERGOB Conference Geneva, B. Guggenheim (Ed.), Karger, Basel, 138–44.
138. GRUPP, U. and SIEBERT, G. (1978). *Res. Exp. Med. (Berl.)*, **173**, 261.
139. MUSCH, K. VON, SIEBERT, G., SCHIWECK, H. and STEINLE, G. (1973). *Z. für Ernährungswissenschaft*, Suppl. 15, 3.
140. SIEBERT, G., GRUPP, U. and HEINKEL, K. (1975). *Nutrition and Metabolism*, **18**, Suppl. 1, 191.
141. KARLE, E. J. and GEHRING, F. (1978). *Dtsch. Zahnärztl. Z.*, **33**, 189.
142. KARLE, E. J. and GEHRING, F. (1979). *Dtsch. Zahnärztl. Z.*, **34**, 551.
143. AMINOFF, C. (1974). In: *Sugars in Nutrition*, H. L. Sipple and K. W. McNutt (Eds.), Academic Press, New York, 135.
144. MALSUO, T. (1973). *Shigaku*, **60**, 760 (*Chem. Abs.* 81,58745, 1974).
145. MAKI, Y., MATSUKUBO, T. and TAKEUCHI, M. (1980). *J. Dent. Res.*, **59**, Special Issue B. IADR Abstract No. 156, 154.
146. IGARASHI, K., KAMIYAMA, K. and YAMADA, T. (1980). *J. Dent Res.*, **59**, Special Issue B. IADR Abstract No. 181, 179.
147. IKEDA, T. *et al.* (1977). *J. Dent. Res.*, **56**, AADR Abstract No. 16, 55.
148. YAMADA, T., KIMURA, S. and IGARASCHI, K. (1980). *Caries Res.*, **14**, 239.
149. CORNICK, D. E. R. and BOWEN, W. E. (1972). *Archs. Oral Biol.*, **17**, 1637.
150. MACDONALD, I. (1966). *Amer. J. Clin. Nutr.*, **18**, 639.

# Chapter 4

# MEDICAL IMPORTANCE OF SUGARS IN THE ALIMENTARY TRACT

I. S. MENZIES

*Department of Chemical Pathology and Metabolic Disorders,
St. Thomas's Hospital Medical School, London, UK*

## SUMMARY

*Recent history has seen a dramatic change in the availability of sugar for human consumption. Formerly an expensive luxury available to the privileged few, it has now become cheap and is produced commercially on a vast scale. At the molecular level, sucrose (cane or beet sugar) contains linked molecules of fructose and glucose, which are excellent sources of metabolic energy. Unfortunately, when taken in excessive quantities, it predisposes to obesity, dental caries and, perhaps, other undesirable medical conditions. Substitution by less harmful 'sweeteners' is a logical method of reducing this modern health hazard. Several alternative sugars and sugar alcohols are poorly assimilated and are of sufficient sweetness to be considered suitable for this purpose, but they tend to produce side effects because they remain unabsorbed within the intestine. This chapter describes the medical importance of sugars in relation to the alimentary tract, in particular the circumstances that predispose to sugar malabsorption, and the consequences, which include both desirable and undesirable features. The use of sugar indicators to assess intestinal absorption, permeability, disaccharide hydrolysis and other features of diagnostic and research interest is also discussed.*

89

## 1. HISTORICAL ASPECTS

Medical interest in the use of sugars for diagnostic, therapeutic and research purposes has encompassed problems in nutrition, metabolism and intestinal function but the study of glucose metabolism and its major defect, diabetes mellitus, has understandably predominated.

The late-19th and early-20th centuries saw enthusiastic attention given to the roles of a wide range of sugars in human physiology and disease. Insight gained at this time was often astonishing, mainly due to ingenious modification and energetic use of the non-specific analytical techniques then available.[1]

Following the isolation of insulin by Banting et al.[2] in 1922 when the role of glucose in diabetes mellitus became more firmly established, interest in the non-glucose sugars underwent a decline, in part due to the lack of specific techniques for analysis. Except for the recognition of 'transferase-deficiency' galactosaemia in 1935,[3] the study of galactose tolerance in liver disease and thyrotoxicosis,[4-6] and the investigation of sucrose and inulin for the measurement of extracellular space and renal clearance,[7,8] most attention during the next 40 years was directed towards the physiology and pathology of glucose.

Further developments took place after 1958 when several authors described a relationship between sugar malabsorption and the onset of intestinal symptoms such as diarrhoea,[9] thus precipitating a revival of interest in 'intestinal sugar intolerance' which had originated some 50 years beforehand. Elucidation of the mechanisms responsible for intestinal sugar hydrolysis and absorption conducted at the same time by Auricchio,[10] Dahlqvist,[11,12] Crane[13] and others laid a good foundation for the subsequent rapid growth in clinical interest.

A study of the effects produced by non-absorbable dietary carbohydrate, especially in relation to intestinal lactose hydrolysis in different human ethnic groups, now became an important extension of the subject. For many years the relevance of dietary fibre in health and in the incidence of certain diseases had received much attention. Wide-ranging claims relating 'unavailable' dietary polysaccharide residues ('fibre') to disease have become popular, but this subject continues to generate more questions than answers. Introduction of lactulose as a laxative during the late 1960s focused attention on poorly absorbed sugars such as the raffinose oligosaccharides which are quite abundant in many vegetable-rich diets. Poorly absorbed dietary constituents, whether polysaccharides, such as cellulose fibre or inulin, or soluble sugars and polyols, such as raffinose, mannitol, sorbitol and xylitol, have broadly similar effects on the intestine, although reactions to soluble sugars may be more dramatic because an osmotic factor is also present.

During the last 30 years the role of metabolically 'available' sugars has also received increasing attention. In contrast to the times when honey and pure sugar were luxuries, overproduction coupled with popularity now make excessive intake a significant threat to health in modern developed countries.[14] It has proved easier to postulate associations between excessive dietary sugar intake and such important causes of ill health as obesity, hypertension, diabetes mellitus and arterial degeneration than to establish the exact scientific relationships. The connection between eating sugar, especially sucrose, and dental caries is on firmer ground.[15]

## 2.  PHYSIOLOGY OF SUGAR ABSORPTION IN THE HUMAN

Dietary carbohydrates are usually assigned to one of two groups: those which are 'available' because they can be absorbed by the small intestine and then metabolised, and those which are 'unavailable'.

The polysaccharides starch and glycogen, and the disaccharides sucrose, maltose, lactose and trehalose, together with their monosaccharide components glucose, galactose and fructose, are all efficiently absorbed and metabolised. These are 'available' carbohydrates that provide an important source of energy in the human diet. Following ingestion, polysaccharides and disaccharides require conversion to monosaccharides before effective absorption can take place; starch and glycogen are hydrolysed to maltose, maltotriose and α-dextrins by the action of pancreatic and salivary α-amylase, while maltose, maltotriose, α-dextrin, sucrose, lactose and trehalose are converted to monosaccharides by specific hydrolytic enzymes, the disaccharidases, which are active at the brush-border surface of the epithelium of the small intestine.[16] These monosaccharides are then taken up rapidly and transported across the epithelial cell layer— glucose and galactose by active mediation (a sodium-dependent energy-requiring step), and fructose by a separate passive mediated process.[17,18]

### 2.1. 'Unavailable' and Poorly Absorbed Carbohydrates
A major portion of plant carbohydrate material cannot be absorbed efficiently by the unaided mammalian intestine. This is either because

hydrolytic conversion to monosaccharide does not take place, or because their monosaccharide constituents have little or no affinity for the existing transport mechanisms. Thus polysaccharides such as cellulose and inulin cannot be hydrolysed by intestinal α-amylases, and oligosaccharides of the raffinose group[19,20] resist the action of brush-border disaccharidases. Monosaccharides such as mannose and sorbose, and the sugar alcohols sorbitol and xylitol have little affinity for mediated transport mechanisms and therefore diffuse very slowly across the intestinal absorptive surface with only 10–30% of the ingested amount being absorbed. Though of limited availability by the oral route, these sugars are metabolised if they reach the circulation but they may produce toxic side-effects when given in large amounts intravenously.[21–3] Mannitol and L-rhamnose are neither absorbed by mediated transport nor metabolised after reaching the circulation while the synthetic glucose analogue 3-$O$-methyl-D-glucose, although actively absorbed, is not metabolised and is excreted unchanged. These monosaccharides are all regarded as 'unavailable' for metabolism when given by mouth but, as explained, the reasons for this vary. Like inulin and sucrose[7,8] most disaccharides and oligosaccharides are not metabolised if they reach the human circulation intact, and are rapidly and completely excreted in the urine when administered intravenously, although maltose is an exception.[24]

Certain portions of the mammalian gut, such as the rumen of cattle and the human colon, function as 'fermentation chambers' where non-digestible dietary components are converted by bacterial activity to products that can be absorbed from the lower intestine.[25] Short-chain fatty acids derived from carbohydrate residues are known to be assimilated by the human colon,[26,27] and to this extent sugars that fail to be absorbed from the small intestine may yet become available for nutritional purposes.

## 3.  MEDICAL ASPECTS OF INTESTINAL SUGAR ABSORPTION

Sugars are of medical interest in relation to the alimentary tract for several reasons: the ingestion of poorly absorbed sugars may alter intestinal behaviour and produce symptoms; the administration or restriction of appropriate sugars can be of value in the treatment and possibly in the prevention, of disease; and certain sugars are employed as indicators for the clinical investigation of intestinal function and integrity.

## 4.  SUGAR MALABSORPTION

Intestinal malabsorption of carbohydrates may arise as a consequence of defective disaccharide hydrolysis or defective monosaccharide transport.[28-30] Either can be the result of a genetically determined ('primary') defect, intestinal damage due to infection (bacterial, protozoal or viral gastroenteritis), coeliac disease (intestinal sensitivity to the gluten of wheat in susceptible individuals), or some other acquired condition ('secondary'). Genetically determined defects of sugar absorption are usually specific, i.e. they affect a single factor only, whereas an acquired disease tends to produce an impairment of several different functions simultaneously, although some (such as lactose hydrolysis) are more vulnerable than others (e.g. glucose transport).

### 4.1.  Lactose Malabsorption
Of the three enzymes with $\beta$-galactosidase activity that have been isolated from the small-intestinal absorptive epithelium, only one, with a pH optimum of 6 and located in the enterocyte brush-border, is responsible for the hydrolysis of ingested lactose—this is referred to as 'lactase' in the following discussion.[31]

Impaired intestinal hydrolysis of lactose is usually a predominant feature of combined disaccharidase deficiency in acute infectious gastroenteritis and coeliac disease in both adults and children. Some newborn infants develop chronic acid diarrhoea and failure to grow when given milk, but thrive on a lactose-free diet, a rare occurrence which appears to be due to a congenital, possibly genetically determined, deficiency of intestinal lactase.[28,32]

### 4.1.1.  Primary Adult Lactase Deficiency
This may be the commonest inherited human enzyme deficiency. In most mammals, intestinal lactase activity is high at birth but declines to very low levels after weaning, a sequence that accords well with the pattern of milk intake and therefore the requirement for lactose hydrolysis. There are some interesting exceptions to this pattern of behaviour, for example, Pinnipedia such as the California sea lion which produces a milk devoid of sugars and lacks intestinal lactase at birth, and human milk-drinking communities in which a high proportion of individuals retain intestinal lactase activity throughout adult life.

Adult intestinal lactase deficiency was first demonstrated in patients with diarrhoea and abdominal symptoms aggravated by milk consumption.

Many of these patients had evidence of damage to the intestinal absorptive epithelium, and were therefore considered to have a secondary lactase deficiency—however, an increasing number revealed no abnormality other than the (primary) enzyme defect. This at first gave rise to the exciting speculation that the cause of a symptom complex known as the 'irritable colon' syndrome had at last been discovered.

Further studies revealed that the majority of human ethnic groups follow the characteristic mammalian behaviour with high lactase levels at birth followed by a fall after weaning, reaching very low levels of activity between three and five years which persist throughout adult life.[33] Thus, 'primary adult lactase deficiency', or 'hypolactasia', is the normal state for most human racial groups. It is seldom associated with the symptoms of sugar malabsorption (see Fig. 1, p. 97) provided that ingestion of fresh milk remains below about 300 ml (equivalent to 15 g of lactose) at any one time. Symptoms of milk intolerance often appear in hypolactasic individuals, however, following the development of inflammatory bowel diseases, especially colitis, or after surgical removal of the stomach.

Habitually milk-drinking communities, such as North European Caucasians and African tribes of Hamitic origin,[34,35] maintain a high level of intestinal lactase activity in adult life with no decline after weaning. After some controversy it became clear that intestinal lactase activity was determined genetically rather than by the personal milk-drinking habits practised during the early life of an individual. Although prolonged feeding of lactose to hypolactasic individuals is often accompanied by an improvement of the initial symptoms of intolerance, no increase in intestinal lactase activity can be demonstrated.[36] Conversely, withdrawal of lactose from the diet for many years, as is undertaken for the treatment of galactosaemia, is not followed by any reduction of intestinal lactase activity.[37] This contrasts with the rapid alteration in sucrase and maltase activity induced by variations in the dietary intake of sucrose and fructose.[38]

Adult lactase production, although a biologically unique phenomenon in mammals, is inherited as an autosomal dominant factor,[39] sometimes referred to as the 'ALP' gene. The occurrence of ALP in present-day human populations varies widely; Caucasian communities in Northern Europe have approximately 10% incidence of adult hypolactasia (80% ALP), Orientals, Australian aborigines and North American Indians have about 90% incidence (10% ALP), while other races range between these two extremes.[33] The distribution of ALP in human populations has stimulated speculation concerning the nature of its induction. Are we to consider, as

McCracken has,[33] that the lactase-deficient individual is at a disadvantage in situations where lactose constitutes a significant portion of the adult diet? Small-intestinal hurry induced by unhydrolysed lactose might, for instance, impair the absorption of milk nutrients. But many Mediterranean, Indian, Iranian and African communities have a high prevalence of hypolactasia even though they enjoy a substantial intake of dairy produce. These products are often subjected to bacterial action before consumption yet surprisingly large amounts of lactose survive and are found in yoghurt, buttermilk and other fermented milk products.[40-2] Flatz and Rotthauwe[43] have suggested that the promotion of calcium absorption associated with the ability to hydrolyse lactose and to assimilate the monosaccharide products might present a selective advantage which favours the appearance of *ALP* in the milk-drinking populations of Northern Europe where low ultraviolet irradiation impairs the synthesis of vitamin D. If so, the high incidence of hypolactasia in Lapps living on the shore of the Arctic Ocean becomes easier to understand. Isokoski *et al.*[39] have suggested that this may be related to the rich supply of vitamin D derived from fish in coastal areas which would maintain calcium absorption and reduce the necessity for the alternative lactose mechanism.

## 4.2. Sucrose Malabsorption

Isolated deficiency of intestinal sucrase was first described by Weijers *et al.*[44] in 1961, and an association with isomaltase ($\alpha$-dextrinase) deficiency was soon discovered.[28] Amongst Greenland Eskimos, who have a diet habitually low in starch, sucrose and lactose, there is a substantial prevalence of intestinal disaccharidase deficiency (54 % adult hypolactasia and 10·5 % congenital asucrasia).[45] Otherwise sucrase–isomaltase deficiency is a very rare occurrence and appears to be due to the inheritance of an autosomal recessive factor affecting the production of a single brush-border protein containing sub-units with both sucrase and isomaltase activity.[16,46] Although the deficiency is thought to be present at birth, the symptoms of fermentative diarrhoea and failure-to-thrive, which can be controlled by restricting sucrose intake, do not appear until this sugar is added to the infant diet.

Certain complex oligosaccharides of microbial origin that competitively inhibit the hydrolysis of $\alpha$-glucosides in the intestine[47] have recently received active attention. A constituent of this group ('acarbose', Bay. g 5421) shows very pronounced inhibitory action against intestinal sucrose hydrolysis and has been prepared for experimental purposes by the Bayer Company. Apparently free from side-effects, except those associated with temporary inhibition of intestinal hydrolysis of starch and sucrose, this substance may provide a useful means for investigating sugar malabsorption and also for restraining carbohydrate assimilation in the treatment of conditions such as diabetes mellitus, obesity and hyperlipidaemia.[48,49]

### 4.3. Glucose–Galactose Malabsorption

This interesting but very rare condition was first described by Lindqvist and Meeuwisse[50] in 1962. Active small-intestinal transport of D-glucose and D-galactose is impaired but fructose absorption is unaffected, a finding which nicely demonstrates the independence of these two pathways. The defect is inherited as an autosomal recessive factor and produces severe acid diarrhoea leading to dehydration and electrolyte deficiency with an onset shortly after birth. On hydrolysis, most dietary carbohydrates yield either glucose or galactose which accumulates within the intestine and produces a persistent severe osmotic diarrhoea. Exclusion of all carbohydrate except fructose from the diet is the only effective treatment, and when implemented leads to rapid and lasting improvement. Excretion of small amounts of glucose in the urine may indicate the presence of an associated mild defect of glucose transport in the renal tubule as well.[51,52]

## 5.  GASTROINTESTINAL EFFECTS PRODUCED BY CARBOHYDRATE ADMINISTRATION

### 5.1. Oral Glucose-Electrolyte Therapy for Acute Diarrhoea

In cholera and other acute diarrhoeal illnesses the correction of fluid and electrolyte depletion is often a life-saving priority. Traditional correction of these losses by the intravenous route often poses serious practical problems related to the cost and availability of suitable equipment, solutions and skilled personnel under field conditions.[53] The demonstration that the presence of glucose in orally administered saline solutions will potentiate intestinal absorption of sodium and water in these circumstances has been claimed as potentially the most important medical advance of the century.[54] It has certainly proved to be a simple and effective substitute for intravenous therapy. More recently, sucrose and starch (as rice powder), which are both more readily available in underdeveloped countries than glucose, have also been found to be as effective as glucose for this purpose.[55]

**5.2. Effects Produced by Unabsorbed Carbohydrate: Sugar Diarrhoea[56]**
The consequences of ingesting poorly absorbed carbohydrate have become the subject of much speculation and research. The following aspects require discussion: the acute laxative effect; the biochemical consequences of sugar fermentation within the alimentary tract; and the longer-term effects related to dietary carbohydrate residues ('dietary fibre').
Diarrhoea is the main feature of a symptom complex (Fig. 1) produced by ingestion of non-absorbable sugar in sufficient quantity. As previously

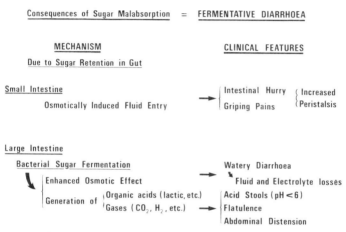

FIG. 1.    The consequences of sugar malabsorption.

mentioned, this situation can arise following the ingestion of oligosaccharides or monosaccharides for which specific intestinal hydrolytic or transport facilities do not normally exist, or following the ingestion of normally absorbable sugars by an individual with defective intestinal hydrolysis or transport.

In practice, few naturally occurring foods contain sufficient concentrations of unabsorbable sugar (e.g. pentoses such as arabinose and xylose, vegetable oligosaccharides such as raffinose, stachyose and verbascose,[19,20] and sugar alcohols such as mannitol, sorbitol and xylitol) to cause overt diarrhoea, although minor symptoms such as abdominal distension and flatulence often arise.[19] Artificially prepared foods, drink and medicines, however, sometimes do contain sufficient poorly absorbed sugars or sugar alcohols (such as sorbitol or xylitol) to cause diarrhoea.[23,57,58] Furthermore, lactulose, a synthetic disaccharide which resists hydrolysis in the human intestine,[59] is marketed by several pharmaceutical firms specifically for use as a laxative.

Several factors (Fig. 1) combine to produce the syndrome referred to as 'sugar diarrhoea'. Of these, 'osmotic purgation' due to osmogenic retention of fluid within the gut in both small and large intestines[60] is probably the most important. Bacterial degradation of unabsorbed sugar entering the colon contributes some 'fermentative' features, such as flatulence and the stool acidity, which are not characteristic of purely osmotic purgatives such as magnesium and sodium sulphates (Epsom and Glauber's salts, respectively), and may provide an additional factor predisposing to diarrhoea.[61,62]

Tolerance to the ingestion of poorly absorbed sugars, expressed as the relationship between dose and symptoms, shows considerable individual variation. Furthermore, the same individual can rapidly develop resistance to the symptoms on repeated oral intake, even though absorption of the sugar remains unchanged.[36] The compensatory mechanisms involved deserve investigation. Sugar diarrhoea may be a nuisance to adults but for infants with acute gastroenteritis it is a life-threatening complication which predisposes to salt and water depletion and demands exclusion of disaccharide from the diet. However, it would be wrong to assume that the effects of unabsorbable sugar are always detrimental: lactulose has become a widely employed and useful purgative, and milk is often taken by hypolactasics for the same purpose; fermentation of unabsorbed sugar in the colon has beneficial effects in the treatment of advanced liver disease; and unabsorbed soluble sugars come within some definitions of 'dietary fibre'[63] and may well share some of its claimed salutary properties (see Section 5.5).

### 5.3. Bacterial Degradation of Carbohydrate in the Intestine

The variation in bacterial activity found in different parts of the alimentary tract is determined by local conditions. The factors involved are not yet well understood[64] but the outcome must depend upon a balance between influx of substrate from both diet and host, and inhibitory influences such as secretion of gastric acid, proteolytic and bacteriolytic enzymes (such as lysozyme), antibodies and bile salts, as well as the efficiency with which contents are cleared along the bowel. Under normal circumstances, bacterial activity is suppressed in the human stomach and small intestine but increases in the oral cavity and especially in the caecum and colon. Abnormalities that disturb the regular flow of contents through the small intestine predispose to bacterial growth in this area, a condition known as the 'blind loop' syndrome which often causes malabsorption.[65] Sugars habitually included in the diet are utilised by bacteria in the oral cavity and, if they fail to be absorbed from the small intestine, are degraded in the large intestine. Both these have consequences of medical interest.

Carbohydrates that evade the small-intestinal absorptive mechanisms pass to the large intestine where they become available for degradation by colonic bacteria. Examples include: so-called 'unavailable' polysaccharides such as cellulose, gums, pectins and inulin and the raffinose oligosaccharides of vegetable origin;[20] poorly absorbed sugars and sugar alcohols used as food additives (sorbitol and xylitol) or in therapy (lactulose); and normally 'available' sugars when they fail to be absorbed as a result of primary or secondary intestinal defects.

Medical interest in these malabsorbed carbohydrates is related to the importance of 'dietary fibre',[63,68] the therapeutic uses of lactulose,[69,70] and, as has already been discussed, sugar diarrhoea as a complication of intestinal disease.[56] Although the different carbohydrates show considerable variation in their susceptibility to degradation by colonic bacteria, this process plays a part in each of these situations.

### 5.3.1. Lactulose Fermentation in the Colon

Degradation of the lactulose by colonic bacteria provides a practical opportunity for the study of sugar fermentation within the large intestine. Weber[70] has recently written a useful review of the subject with reference to the employment of lactulose for the treatment of portal–systemic encephalopathy. Lactulose (4-$O$-$\beta$-D-galactopyranosyl-D-fructo-furanose) is a keto analogue of lactose that was first synthesised from lactose by the 'Lobry de Bruyn–Alberda van Elkenstein' transformation (alkaline isomerisation) of glucose to fructose within the disaccharide molecule.[71] Unlike lactose, it resists the hydrolytic action of small-intestinal $\beta$-galactosidases,[59] remains largely unabsorbed and enters the colon. Although in this respect it does not differ in behaviour from the raffinose oligosaccharides, nor indeed from lactose when this sugar has been ingested by a hypolactasic individual, lactulose has come to be regarded as the standard agent for achieving entry of sugar into the large intestine for therapeutic purposes.

The products of lactulose fermentation by colonic bacteria include lactic acid and smaller amounts of acetic and formic acids,[72] although butyric, propionic and other organic acids may also appear in amounts that depend upon the exact conditions and bacterial species present.[73] Gases, mainly carbon dioxide, hydrogen and methane, appear[74] and are a source of some anxiety because of the risk of explosions, which may be fatal, during bowel surgery.[75-7] Production of organic acids renders the stool characteristically acid, and there is evidence that although these products are partly absorbed[26,27] they may also irritate the colon[61,62] and enhance the osmotic effect of lactulose[78] initiated in the upper intestine.[60] All of these effects tend to accentuate the purgative action of lactulose.

*5.3.2. Lactulose in the Treatment of Portal–Systemic Encephalopathy*[70]

Portal–systemic encephalopathy (PSE) is a term used to describe the neurological effects produced when circulating levels of ammonia and other toxic nitrogenous products, derived from bacterial degradation of nitrogen-containing substrates within the intestine, rise because of defective hepatic clearance. These effects vary from mild personality changes to frank coma and can be improved by reducing dietary protein intake or suppressing bacterial activity within the intestine. In either case, the intention is to reduce the 'proteolytic' conversion of nitrogen-containing substrates to ammonia within the gut, but this is achieved at the cost of accentuating protein depletion in the first instance and the risk of unwanted complications of prolonged antibiotic treatment in the second.

Purgative treatment, which reduces the time available for the generation of ammonia etc. and assists with the elimination of toxic metabolites by speeding the passage of intestinal contents, is also considered of value. However, the suggestion that lactulose might benefit patients with PSE, attributed to Ingelfinger[79] in 1965, was based on Metchnikoff's idea that conversion of the bacterial population of the large intestine to a mainly 'saccharolytic' flora dominated by lactobacilli might produce a beneficial reduction in the 'proteolytic' generation of toxic amines. The value of this suggestion was soon confirmed—Bircher et al.[80] found lactulose to be more effective than the antibiotic neomycin for correcting PSE. Watery diarrhoea, the only side-effect, was produced when a high dose was given (optimal range 70–100 g lactulose daily).

The action mechanism of lactulose in the treatment of PSE has proved more complex than that suggested by Ingelfinger.[79] The growth of colonic lactobacilli can be potentiated by lactulose administration, but the beneficial effects precede this change and also occur in the absence of clinical purgation. At the low pH induced by lactulose fermentation ammonia becomes converted to ionic $NH_4^+$ which permeates the colonic wall less efficiently than non-ionic ammonia; the occurrence of such 'ammonia trapping' has not been supported by the demonstration of increased faecal ammonia content.[81] However, there is good evidence that accentuation of bacterial growth induced by the provision of substrate in the form of lactulose is associated with an increased conversion of nitrogen available within the colon to bacterial protein and a decreased generation, and possibly some utilisation, of ammonia.[82] Similar effects are likely to result when other poorly absorbed carbohydrates are administered, although dose–response relationships would depend upon relative absorption from the small intestine and the affinities of colonic bacteria for the administered sugar.

Lactulose therapy has also been tried in the management of uraemia, dysentry and dental disease but without the obvious beneficial result seen in the case of constipation and PSE.[69]

### 5.4.  Sugars, Oral Bacteria and Dental Caries

A relationship between repeated exposure of dental surfaces to sugar and the production of caries has been clearly established.[15] Freedom from dental caries is a conspicuous feature of patients with hereditary fructose intolerance who avoid sucrose and other dietary sources of fructose during their whole lifetime and indicates that these sugars are particularly active in this respect. Repeated exposure of certain species of bacteria to particular types of sugar (designated 'cariogenic') within the oral cavity leads to the formation of a gelatinous matrix of bacteria surrounded by polysaccharide, called 'plaque', on the dental surfaces. Plaque forms the basis of an ecosystem which facilitates the degradation of sugar to acidic products with dental erosion leading to caries and associated complications.[66]

Research programmes, such as the Turku sugar studies,[67] intended to define predisposing factors and establish details of the cariogenic process form a valuable basis for the control of this important disease. One measure in preventive treatment involves inhibition of microbial sugar metabolism and plaque formation by avoiding cariogenic sugars, as discussed in Chapter 3.

### 5.5.  Importance of Fibre and Other Carbohydrate Residues[68]

The polysaccharides cellulose and hemicellulose, partly in combination with the phenyl–propane polymer lignin, form the true structural fibre of vegetable tissue. The term 'dietary fibre' has been used to describe the 'residue of plant cells left after digestion by the alimentary enzymes of man'.[63]

A good, if relatively early, review of the subject of dietary fibre and its medical importance appeared in 1976.[68] Burkitt,[63] an active proponent of the 'high-fibre diet', has put forward 'a hypothesis attempting to explain the epidemiological features and the inter-relationships of certain characteristically Western diseases' which 'postulates that fibre-depleted diets may play a role in the cause of each disease without necessarily being the sole cause of any'. He proposes that a link exists, in order of diminishing strength, between dietary fibre depletion and: (1) constipation; (2) diverticular disease of the colon; (3) appendicitis, haemorrhoids and hiatus hernia; and (4) varicose veins and bowel cancer.

By increasing stool bulk and softening stool consistency, high dietary fibre intake increases the speed of bowel transit thereby preventing constipation and a number of associated conditions.[83] Western diets low in fibre may predispose to constipation with straining and frequent rises in abdominal pressure. This, in turn, predisposes to the development of haemorrhoids, varicose veins and hiatus hernia.

Diverticular disease (inflammation of sac-like protrusions through the muscle layer of the colon) is associated with raised pressures within the colonic lumen. Low-residue diets tend to raise intracolonic pressure, and bran supplements have been shown to reduce colonic pressure and alleviate symptoms.[84,85]

The connection between dietary fibre and bowel cancer is more complex.[86] The main issue concerns the ability of dietary fibre to reduce exposure of the bowel to carcinogens. Some of these are believed to be generated from precursors such as bile acids by bacterial activity within the colon.[87] Dietary fibre is thought to inhibit the synthesis of such carcinogens by binding the precursors and modifying bacterial metabolism, and, furthermore, it might reduce bowel exposure to these products by adsorption, dilution and speeding their clearance.[83]

The bile-salt-binding properties[88] and epidemiological associations[89] of dietary fibre suggested that it might also be capable of influencing cholesterol metabolism and altering the incidence of atheromatous degeneration of arteries (atheroma) and gallstone disease.[90] High fibre diets are known to increase the stool excretion of bile acids, lower serum cholesterol levels and prevent the development of atheroma in experimental animals such as the rabbit.[91] Under normal circumstances, bile salts enter the duodenum by the common bile duct and are largely reabsorbed from the ileum to be re-excreted by the liver in the bile, which helps to conserve bile salts. Dietary fibre, especially pectin and guar gum, interrupts this cycle by binding and preventing the reabsorption of bile salts. This necessitates a compensatory increase in hepatic bile-salt synthesis from the cholesterol pool, and may also impair intestinal absorption of cholesterol by inducing a bile-salt deficiency—both effects tending to lower blood cholesterol concentration. There are several reports that 'fibre' supplementation of the diet, especially with pectin and guar gum, can lower human blood cholesterol levels,[92-4] and an inverse correlation between the level of the fibre in the diet and the incidence of human coronary heart disease has been noted by one group of investigators.[95]

A suggestion that refined Western diets with a low fibre content might encourage the development of diabetes mellitus,[96,97] and the need to control atheroma which is a common serious complication, has led to several trials of fibre-supplemented diets in this condition. Inclusion of pectin, guar gum[98] and wheat bran[93] improved the control of diabetes mellitus, as indicated by reductions in urinary glucose excretion and the dose of insulin required. It is not clear whether this effect is due to lower intestinal sugar absorption or to an improvement in endogenous insulin production.

As previously defined, dietary fibre has been held to include all the 'unavailable' carbohydrates of the diet.[63] In addition to 'crude fibre' this includes constituents such as pectin, gums and the raffinose oligosaccharides which are by no means 'fibrous'. The term 'dietary residue' which can be qualified as being 'fibrous', 'carbohydrate', 'insoluble', 'soluble', etc., as required, is less misleading and removes any temptation to use the extraordinary expression 'liquid fibre' to describe the soluble fractions. Attention should be drawn to the wide range and abundance of soluble carbohydrate residues in many human diets. Besides the raffinose oligosaccharides in beans, peas and lentils,[19,20] there is lactulose in most processed milk products,[71] and sorbitol, mannitol and xylitol which are often added to commercially prepared foods, medicines and confectionery.[21,58] Lactose should also be regarded as a soluble dietary carbohydrate residue in communities with a high incidence of hypolactasia, whether of the 'primary adult' or 'secondary' type, and may be of considerable importance when fresh or fermented dairy products[40-2] are imported or habitually prepared for consumption. Effects that have been claimed for dietary fibre may equally well apply to these soluble carbohydrate residues.

Much is owed to those who, like Burkitt, have so effectively stimulated medical science with the hypothesis of 'dietary-fibre depletion', but many more years may yet be required before a true assessment of this subject is obtained.

## 6.  USE OF SUGARS FOR EVALUATING INTESTINAL FUNCTION

Crane[99] has reviewed the early development of methods for measuring intestinal transport and absorption of sugars in experimental animals and man. There are different complementary approaches to this subject. On the one hand there are procedures *in vitro* which use isolated tissues and therefore allow greater anatomical precision and selective control of experimental conditions; here the experimental environment is intentionally simplified but extrapolation of the findings to physiological and pathological reality must be undertaken with caution. Methods *in vivo*, on the other hand, have a closer relevance to physiological and clinical situations but are often difficult to interpret because of their anatomical and biochemical complexities.

### 6.1.  Procedures *In Vitro*
Detailed investigation of the mechanisms of digestion and absorption, especially for cellular location, may be difficult or impossible to undertake *in vivo* and yet quite easily performed using viable animal or human tissues *in vitro*. Sections of whole intestine, either cannulated[100] or secured as closed inverted sacs[101] were used for this purpose at first; later, more selective methods using isolated intestinal brush-border, as fragments[102] or artificially formed into vesicles[103] or other subcellular fractions[104,105] were developed. Incorporation of cellular constituents into black lipid membranes or liposomes[106] is an ingenious recent technical development that permits certain enzymes and carrier proteins to be studied in isolation.

### 6.2.  Procedures *In Vivo*
Aspects of intestinal function, such as absorption, can often be effectively, although indirectly, assessed by measuring the rise in blood concentration or renal excretion of a selected test substance following the oral administration of a standard dose. This is one form of the 'function test', a well-established principle of clinical investigation.

Information of greater anatomical specificity can be obtained by delivering test solutions at a defined level in the intestine, or by perfusing a selected intestinal segment by means of an oral or nasal tube. The Nicholson–Chornack intubation technique[107] was a useful early modification of the tube method which was originally introduced by Miller and Abbott[108,109] in 1934.

High-molecular-weight polyethylene glycol (PEG), first used in veterinary studies,[110] was introduced in 1957 by Borgström *et al.*[111] for use in human intestinal perfusion as a non-absorbable indicator to overcome the effects of bi-directional fluid exchange and facilitate calculations of absorption. PEG has since been widely used with various modifications to the perfusion technique to assess intestinal absorption of sugars[112–16] and other constituents. The value of using a second marker to indicate unmediated diffusion and allow for variables, such as concentration gradient and the time and area of exposure to the absorptive surface suggested by Fordtran,[117] has recently been demonstrated by using urea to monitor these diffusion factors collectively.[118]

### 6.3. Intestinal Function Tests Used in Clinical Medicine
Sugars have been used as indicators for clinical function tests that evaluate absorption, permeability, disaccharide hydrolysis and the speed of intestinal transit, all of which are features that assist the investigation of intestinal integrity.

### 6.3.1. Mediated Absorption
The 'glucose-tolerance test' has been applied mainly to the problem of diagnosing diabetes mellitus but it was formerly also regarded as a test of intestinal absorption. The maximum (peak) rise in blood glucose level following oral administration of this sugar is reduced in the presence of malabsorption. Although the difference in group response between patients with malabsorption and healthy subjects is of high statistical significance, individual responses show considerable overlap between the two groups with poor diagnostic discrimination.[119] As an indicator of intestinal absorption glucose has three main disadvantages: it is continuously present in the circulating blood, and the concentration may fluctuate for metabolic reasons; behaviour after ingestion is influenced by several factors unrelated to intestinal absorption; and absorption of glucose is so efficient that it is insensitive to minor defects of intestinal transport. However, the test remains useful for differentiation from pancreatic disease which often produces both malabsorption and glucose intolerance resulting from exocrine and islet-cell defects, respectively.

The D-xylose-absorption test, introduced in 1937 by Helmer and Fouts,[120] was first employed for the investigation of clinical malabsorption by Fourman in 1948.[121] It has since become established as a test of intestinal absorptive capacity but is used in a variety of forms. D-Xylose is poorly metabolised in comparison with D-glucose, and is therefore less affected by factors other than absorption. The choice of optimal test dose has varied considerably. For adults some authors employ $25\,g^{122-8}$ while others use 15[129] or 5 g[130-2] on the grounds that the larger dose may produce intestinal hurry and osmotic purgation. Infants and children receive either a fixed 5 g dose[133] or a variable dose related to body weight[134-8] or surface area.[139] Interpretation is based upon the behaviour of the blood xylose concentration/time (or 'absorption') curve, or timed excretion in the urine, but the influence of factors other than the state of intestinal absorption on these measurements presents a problem.

The advantages of a non-invasive clinical test involving ingestion of an indicator followed by collection of blood or urine samples are simplicity of performance and acceptability to the patient. Variations depending on the large number of factors that influence this type of test can be overcome only by careful selection of the indicator and detailed attention to procedure. The D-xylose test is influenced by variations in gastric emptying rate; in addition, blood concentrations will be influenced by body size, and recovery in the urine by the state of renal clearance and the competence of sample collection. Interpretation based on blood xylose concentrations at 60 min is unaffected by poor renal clearance, whether due to old age[140] or renal disease. Correction to a constant body surface area should overcome the disadvantages of calculations based on body size.[132]

The potential advantages of 3-O-methyl-D-glucose (3mGlc) for use as an indicator of human intestinal absorption were first investigated by Fordtran et al.[112] in 1962. This synthetic monosaccharide, like D-glucose and D-galactose, is transported actively by the small intestine—unlike them, however, it is neither metabolised nor actively reabsorbed by the renal tubule, with 90 % of a 10–20 g oral dose recovered from human urine within 48 h. Although an ideal indicator for assessing the state of active intestinal hexose transport[141,142] and (unlike D-xylose) completely resisting metabolism and therefore unaffected by variations in hepatic function, 3mGlc was (like glucose) found to be too efficiently absorbed to rival D-xylose as a sensitive indicator of intestinal function.[112]

However, 3mGlc has been combined with D-xylose to provide a differential absorption system which exploits the contrasting effects of impaired intestinal absorption.[143] Plasma D-xylose/3mGlc ratios measured 60 min after oral administration (5·0 and 2·5 g, respectively) gave a more reproducible normal range and better diagnostic discrimination than D-xylose concentration alone. In this procedure, 3mGlc provides a basis for correction for several variables; these include gastric emptying and the space of distribution after absorption. A sensitive thin-layer chromatographic technique is used to overcome the problem of analysing both D-xylose and 3mGlc in the presence of glucose.[144]

*6.3.2. Non-Mediated Intestinal Permeation*

Although intestinal permeability was first studied in the human by Fordtran *et al.*[145] in 1965, their method was based on the estimation of a reflection coefficient and was applicable only to the permeation of molecules below the size of mannitol (molecular weight 180, radius 0·4 nm) which they chose as a non-absorbable reference marker. Later, direct measurements of non-mediated transfer across the intestinal epithelium were made with lipid-insoluble indicators of differing molecular weights by estimating uptake from the intestinal lumen[146] or entry from blood into the intestinal lumen[147] using intubation with perfusion.

Clinical assessment of intestinal permeability became a practical possibility with the introduction of less invasive procedures. The fraction of an orally administered test substance subsequently recovered in the urine may be considered a measure of intestinal permeability provided it is established that the indicator chosen permeates the intestinal wall by unmediated diffusion, and that the absorbed fraction is fully excreted.[24] Most indicators suitable for this purpose are absorbed so slowly that concentrations building up in the blood following an acceptable oral dose are, unlike urine concentrations, too low for satisfactory quantitation.[24]

Lactulose (molecular weight 342, radius 0·5 nm) was suggested for use as an intestinal permeability marker in 1969.[148] Like most oligosaccharides, it resists metabolism and is rapidly and almost completely excreted in human urine following intravenous administration.[24,149] Melibiose, raffinose, stachyose[24,150] and a fluorescein-labelled preparation of the polysaccharide dextran[150] (molecular weights 342, 504, 666 and 3000, radii 0·5, 0·59, 0·62 and 1·25 nm, respectively), like lactulose, resist the action of human intestinal disaccharidases and have been used as permeability markers. Cellobiose, although slowly hydrolysed by intestinal lactase,[12] has also been used for this purpose.[151,152]

Intact oligosaccharides permeate the human intestine very slowly, less than 1 % of the ingested dose appearing in the urine.[24] Intestinal permeability to these relatively large molecules increases when the intestine is damaged, as in coeliac disease,[24,151,153] some types of acute gastroenteritis and, temporarily, following the ingestion of hyperosmolar solutions[24,149,150] or cetrimide.[24] In contrast, the permeation of monosaccharide markers, such as mannitol[151,152] and L-rhamnose,[153] across the healthy intestinal epithelium is much greater (8–30 % of an oral dose), is not increased by ingestion of hyperosmolar solutions, but decreases when the absorptive area is reduced by villous atrophy as in coeliac disease.[151,153] It appears that the permeation of the healthy intestinal epithelium by oligosaccharides takes place through a limited number of large pores, probably associated with the paracellular 'tight junctions', whereas the monosaccharide markers pass mainly through the cells of the mucous epithelium by way of many smaller water-filled pores in the cell walls[145] which oligosaccharides are unable to penetrate.

The 'permeability' monitored by these two groups of markers reflects the state of two different pathways, monosaccharide transfer relating mainly to absorptive area and oligosaccharide transfer relating to the presence of channels that become more numerous when the epithelium is damaged and exposed to a hyperosmolar environment. Both types of permeability can be estimated at the same time by using the markers in combination. Thus both cellobiose–mannitol[151,152] and lactulose–L-rhamnose[153,154] tests of differential permeability have been described. The use of such 'differential permeation ratios' reduces the effect of variations in gastric emptying, renal clearance and urine collection on the outcome of these tests.

The possibility that sugars might be degraded by intestinal bacteria[155] has led to the introduction of markers like PEG 400[156] that resist the action of bacteria. However, intestinal absorption of PEG 400 is 30–100 times greater than that of oligosaccharides of the same molecular weight, and is reduced in coeliac disease but unaffected by hyperosmolar solutions, so that PEG behaves like mannitol and L-rhamnose in these respects. The probable explanation for this is that PEG 400, being more lipophilic than the oligosaccharides, can pass through mucosal cell walls.[150] There are evidently several different mechanisms of intestinal permeability, and at present the oligosaccharide markers are unique in monitoring an aspect of permeation which increases in certain circumstances of disease and diet. There is speculation concerning increased entry of potentially harmful dietary constituents under these circumstances.[149]

### 6.3.3. Disaccharide Hydrolysis

Investigation of the clinical and nutritional importance of defective intestinal disaccharide hydrolysis has depended upon the development of convenient methods of assay. Intestinal perfusion is the most direct and physiological approach,[115,116] but less elaborate procedures are required for routine clinical purposes.

The traditional view that disaccharide hydrolysis takes place within the intestinal lumen by the action of secreted enzymes was questioned by Borgström et al.[111] in 1957 and disproved shortly afterwards following the demonstration of high disaccharidase activity in the brush-border of the small-intestinal epithelium.[11,103,157,158] In 1964, Dahlqvist[12] described a simple method for the assay of disaccharidase activity in intestinal mucosal homogenates, based upon the generation of glucose under controlled conditions of incubation with specific disaccharide substrates. This assay has been widely applied to tissue samples obtained by means of the Crosby capsule[159] or the Rubin intestinal biopsy tube.[160] Modifications of the assay technique, such as the use of umbelliferine-complexed substrates,[104] have also been introduced. Although it is a greater ordeal for the patient than the indirect methods to be described, and relates to a very limited area of intestine, this procedure gives reliable results and offers the additional advantage of histological data.

Several types of indirect test have been introduced; these assess the state of intestinal disaccharide hydrolysis by measuring either the absorption of monosaccharide products, or retention of unhydrolysed substrate within the intestine after oral administration of a standard dose of disaccharide. The reliability of these techniques depends upon the number of uncontrolled factors besides disaccharide hydrolysis that influence the response.

Four versions of the original lactose-tolerance test were published in 1965. They differed with respect to the amount of lactose and volume of water used in the oral test solution (100 g in 250 ml;[161] 100 g in 400 ml;[162] 50 g in an unmeasured volume;[163] and 50 g in 400 ml,[164] respectively), and also the timing of blood sampling. Newcomer and McGill[165] considered that a rise in blood glucose of 20 mg per 100 ml or more above the fasting level following administration of either 50 or 100 g of lactose excluded lactase deficiency, but they stressed that blood samples should be taken with sufficient frequency (15-min intervals during the first hour) to catch the peak maximum. In its unmodified form this test is particularly vulnerable in two respects: first, defective monosaccharide transport often complicates interpretation when disaccharidase deficiency is secondary to intestinal disease; and second, efficient tissue uptake may mask a normal absorption curve and produce a misleadingly poor rise in blood glucose.

The problems which complicate interpretation of a blood glucose absorption curve have been tackled in various ways. By comparing the area under glucose absorption curves following ingestion of: (1) lactose, and (2) an equivalent glucose–galactose mixture, Cuatrecasas et al.[36] calculated an 'absorption ratio' which made allowance for the state of intestinal transport and tissue uptake of glucose. In place of a glucose-absorption curve, Isokoski et al.[166] estimated blood galactose 40 min after lactose ingestion, augmenting the concentration by administering ethanol immediately before lactose administration.[167,168] The demonstration by Cozzetto[169] in 1964 that the normal exhalation of $^{14}CO_2$ in human breath following oral administration of [$^{14}$C]lactose was not observed in a child with lactose intolerance suggested a further indirect method of assessing intestinal disaccharide hydrolysis by monitoring product absorption, which has been considered particularly useful for field work and epidemiological studies.[170-2] Unfortunately, all of these methods depend upon the integrity of intestinal monosaccharide transport, and become unreliable when this is disturbed in certain pathological conditions.

In principle, the remaining methods relate hydrolysis of ingested disaccharide to the persistence of unhydrolysed substrate within the intestine rather than to the absorption of monosaccharide product. Unhydrolysed disaccharide remains largely unabsorbed within the small-intestinal lumen where it induces osmotic influx and retention of fluid, peristaltic hurry and rapid entry into the colon. Laws and Neale[60,173] employed x-ray fluoroscopy to demonstrate dilution and hurry within the intestine following ingestion of a barium sulphate–lactose (or sucrose) mixture as evidence of impaired disaccharide hydrolysis, but found that this response might be suppressed in the presence of villous atrophy. Exhalation of hydrogen derived from bacterial fermentation of un-absorbed carbohydrate entering the colon, which can be conveniently detected by gas–liquid chromatography, forms the basis of several popular 'hydrogen-breath tests' for sugar absorption.[174] This principle has been used to demonstrate impaired intestinal hydrolysis of ingested lactose[175,176] and sucrose,[177] but would become unreliable in the presence of monosaccharide malabsorption.

The amount of intact disaccharide excreted in human urine is closely similar to the quantity permeating across the intestinal wall and entering the blood stream.[24] Intestinal concentrations of lactose and sucrose following ingestion are determined by the rate of hydrolysis, and remain high, predisposing to increased permeation and renal excretion of the intact disaccharide, when hydrolysis within the intestine is impaired.[24] Cook and Howells[178] found that a rise in the excretion of lactose in the urine following oral administration served as a simple screening test for a deficiency in intestinal lactase activity. A further development came in 1974 when Menzies[24,179] found a good correlation between lactase and sucrase activities measured in jejunal biopsy tissue and urinary lactose/lactulose and sucrose/lactulose excretion ratios following oral administration of these disaccharides. In this procedure lactulose, which behaves like intact lactose and sucrose in all respects other than susceptibility to the action of intestinal disaccharides,[59] is employed as an 'internal marker' to make allowance for the effects of gastric emptying, dilution by intestinal secretion, intestinal permeation, renal clearance and other variables that may also influence urinary excretion of ingested disaccharide. Unlike most available techniques, this method is not invalidated by defective mono-saccharide transport, and can be used to investigate, simultaneously, intestinal hydrolysis of lactose and sucrose, and permeability to lactulose in gastrointestinal disease. The urine samples, easily preserved for long periods, can be analysed for disaccharide content by quantitative paper[180] or gas–liquid[181] chromatography.

### 6.3.4. Small-Intestinal Transit and Bacterial Overgrowth

Bond and Levitt[182] established by a tube technique that the appearance of hydrogen in the breath coincided closely with the arrival of a PEG marker in the terminal ileum when both PEG and lactulose had been ingested simultaneously. Evidently bacterial generation of hydrogen is rapid when lactulose has reached the caecum, and the interval between lactulose ingestion and the start of hydrogen expiration can serve as a non-invasive measure of small-intestinal transit time. However, the dose of lactulose should be kept low (5 g) to avoid osmotic hurry. $^{14}CO_2$ in the breath following ingestion of mannitol-1[$^{14}$C][183] or [$^{14}$C]D-xylose[184] has been shown to be a feature of small-intestinal bacterial overgrowth ('blind loop' syndrome).

## 7.  CONCLUSIONS

In conclusion, the last 25 years have seen a revival of interest in the subject of sugar absorption and the medical consequences of both the excessive assimilation and the malabsorption of carbohydrates. The introduction of several new sugar 'markers' with properties appropriate to the measure-ment of different features of intestinal absorption and permeability has been one of the many recent advances in this subject. The use of these markers in combination is a principle with considerable potential,[185] which may give rise to differential absorption tests of clinical value in the future.

## REFERENCES

1. BROCK, H. J. and HUBBARD, R. S. (1935). *Amer. J. Dig. Dis. and Nutr.*, **2**, 27.
2. BANTING, F. G., BEST, C. H., COLLIP, J. B., CAMBELL, W. R. and FLETCHER, A. A. (1922). *Can. Med. Assoc. J.*, **12**, 141.

3.  MASON, H. H. and TURNER, M. E. (1935). *Amer. J. Dis. Childh.*, **50**, 359.
4.  ALTHAUSEN, T. L., LOCKHART, J. C. and SOLEY, M. H. (1941). *Amer. J. Med. Sci.*, **199**, 342.
5.  MACLAGAN, N. F., RUNDLE, F. F., COLLARD, H. B. and MILLS, F. H. (1940). *Quart. J. Med.*, **9**, 215.
6.  MERANZE, D. R., LIKOFF, W. B. and SCHNEEBERG, N. G. (1942). *Amer. J. Clin. Path.*, **12**, 261.
7.  DEANE, N., SCEINER, G. E. and ROBERTSON, J. S. (1951). *J. Clin. Invest.*, **30**, 1468.
8.  DEANE, N. and SMITH, H. W. (1955). *J. Clin. Invest.*, **34**, 681.
9.  Leader: Sugar Diarrhoea (1966). *Lancet*, **2**, 736.
10. AURICCHIO, S. RUBINO, A., TOSI, R., SEMENZA, G., LANDOLT, M. KISTLER, H. and PRADER, A. (1963). *Enzymol. Biol. Clin. (Basel)*, **3**, 193.
11. DAHLQVIST, A. and BORGSTROM, B. (1961). *Biochem. J.*, **81**, 411.
12. DAHLQVIST, A. (1964). *Analyt. Biochem.*, **7**, 18.
13. CRANE, R. K. (1960). *Phys. Rev.*, **40**, 789.
14. AYKROYD, W. R. (1974). In: *Sugar in Nutrition*, H. L. Sipple and K. W. McNutt (Eds.), Academic Press, New York.
15. RUSSELL, A. L. (1974). In: *Sugars in Nutrition*, H. L. Sipple and K. W. McNutt (Eds.), Academic Press, New York.
16. GRAY, G. M. (1981). In: *Physiology of the Gastrointestinal Tract*, L. R. Johnson (Ed.), Raven Press, New York.
17. CASPARY, W. F. (1977). In: *Workshop Hoechst*, Vol. 4, M. Kramer and F. Lauterbach (Eds.), Excerpta Medica, Amsterdam-Oxford.
18. KIMMICH, G. A. (1981). In: *Physiology of the Gastrointestinal Tract*, L. R. Johnson (Ed.), Raven Press, New York.
19. CRISTOFARO, E., MOTTU, F. and WUHRMANN, J. J. (1974). In: *Sugars in Nutrition*, H. L. Sipple and K. W. McNutt (Eds.), Academic Press, New York.
20. KANDLER, O. and KOPF, H. (1980). In: *The Biochemistry of Plants*, Vol. 3, J. Preiss (Ed.), Academic Press, New York.
21. BIRCH, G. G. (1971). *Lancet*, **2**, 1419.
22. MENG, H. C. (1974). In: *Sugars in Nutrition*, H. L. Sipple and K. W. McNutt (Eds.), Academic Press, New York.
23. WANG, Y-M. and VAN EYS, J. (1981). *Ann. Rev. Nutr.*, **1**, 437.
24. MENZIES, I. S. (1974). *Biochem. Soc. Transac.*, **2**, 1042.
25. RERAT, A. (1981). *Reprod. Nutr. Develop.*, **21**, 815.
26. MCNEIL, N. I., CUMMINGS, J. H. and JAMES, W. P. T. (1978). *Gut*, **19**, 819.
27. RUPPIN, H., BAR-MEIR, S., SOERGEL, K. H., WOOD, C. M. and SCHMITT, M. G. (1980). *Gastroenterology*, **78**, 1500.
28. PRADER, A. and AURICCHIO, S. (1965). *Ann. Rev. Med.*, **16**, 345.
29. BAYLESS, T. M. and CHRISTOPHER, N. L. (1969). *Amer. J. Clin. Nutr.*, **22**, 181.
30. GRAY, G. M. (1976). *New Eng. J. Med.*, **294**, 1057.
31. LEBENTHAL, E., TSUBOI, K. and KRETCHMER, N. (1974). *Gastroenterology*, **64**, 1107.
32. HOLZEL, A., SCHWARZ, V. and SUTCLIFFE, K. W. (1959). *Lancet*, **1**, 1126.

33. McCRACKEN, R. D. (1971). *Current Anthropol.*, **12**, 479.
34. SIMOONS, F. J. (1978). *Digest. Dis.*, **23**, 963.
35. COOK, G. C. and KAJUBI, S. K. (1966). *Lancet*, **1**, 725.
36. CUATRECASAS, P., LOCKWOOD, D. H. and CALDWELL, J. R. (1965). *Lancet*, **1**, 14.
37. KOGUT, M. D., DONNELL, G. N. and SHAW, K. N. F. (1967). *J. Pediat.*, **71**, 75.
38. ROSENWEIG, N. S. and HERMAN, R. H. (1969). *Gastroenterology*, **56**, 500.
39. ISOKOSKI, M., SAHI, T., VALLAKO, K. and TAMM, A. (1981). *Annal. Clin. Res.*, **13**, 164.
40. SHANLEY, R. M. (1973). *Austral. J. Dairy Technol.*, **28**, 58.
41. GOODENOUGH, E. R. and KLEYN, D. H. (1975). *J. Dairy Sci.*, **59**, 601.
42. AZADEH, B., ABULNOUR, A., BRAMLEY, P. M. and MENZIES, I. S. (1979). *Proc. Nutr. Soc.*, **38**, 39A.
43. FLATZ, G. and ROTTHAUWE, H. W. (1973). *Lancet*, **2**, 76.
44. WEIJERS, H. A., VAN DE KAMER, J. H., DICKE, W. K. and IJSSELING, J. (1961). *Acta Paediat. Stockh.*, **50**, 55.
45. McNAIR, A., GUDMAND-HØYER, E., JARNUM, S. and ORRILD, L. (1972). *Brit. Med. J.*, **2**, 19.
46. CONKLIN, K. A., YAMASHIRO, K. M. and GRAY, G. M. (1975). *J. Biol. Chem.*, **250**, 5735.
47. SCHMIDT, D. D., FROMMER, W., JUNGE, B., MULLER, L., WINGENDER, W., TRUSCHEIT, E. and SCHAFER, D. (1977). *Naturwissenschaften*, **64**, S535.
48. CASPARY, W. F. (1978). *Lancet*, **1**, 1231.
49. SJOSTROM, L. and WILLIAM-OLSSON, T. (1981). *Current Therapeutic Res.*, **30**, 351.
50. LINDQVIST, B. and MEEUWISSE, G. W. (1962). *Acta Paediat. Uppsala*, **51**, 674.
51. ANDERSON, C. M., KERRY, K. R. and TOWNLEY, R. R. W. (1965). *Arch. Dis. Childh.*, **40**, 1.
52. MARKS, J. F., NORTON, J. B. and FORDTRAN, J. S. (1966). *J. Pediat.*, **69**, 225.
53. Leader: Oral Glucose/Electrolyte Therapy for Acute Diarrhoea (1975). *Lancet*, **1**, 79.
54. Leader: Water with Sugar and Salt (1978). *Lancet*, **2**, 300.
55. MOLLA, A. M., SARKER, S. A., HOSSAIN, M., MOLLA, A. and GREENHOUGH, W. B. (1982). *Lancet*, **1**, 1317.
56. As for Reference 9.
57. GRYBOSKI, J. D. (1966). *New Eng. J. Med.*, **275**, 718.
58. CHARNEY, E. B. and BODURTHA, J. N. (1981). *J. Pediat.*, **98**, 157.
59. DAHLQVIST, A. and GRYBOSKI, J. D. (1965). *Biochim. Biophys. Acta*, **110**, 635.
60. LAWS, J. W., SPENCER, J. and NEALE, G. (1967). *Brit. J. Radiol.*, **40**, 594.
61. BENNETT, A. and ELEY, K. G. (1976). *J. Pharm. Pharmacol.*, **28**, 192.
62. SAUNDERS, D. R. and SILLERY, J. (1982). *Digest. Dis. Sci.*, **27**, 33.
63. BURKITT, D. P. and MENDELOFF, A. I. (1976). *Digest. Dis.*, **21**, 103.
64. GRACEY, M. S. (1981). *Brit. Med. Bull.*, **37**, 71.
65. JONAS, A., KRISHNAN, C. and FORSTNER, G. (1978). *Gastroenterology*, **75**, 791.
66. MAKINEN, K. K. (1974). In: *Sugars in Nutrition*, H. L. Sipple and K. W. McNutt (eds.), Academic Press, New York.
67. SCHEININ, A. and MAKINEN, K. K. (1974). *Acta Odontol. Scand.*, **32**, 383.

68. SPILLER, G. A. and AMEN, R. J. (Eds.) (1976). *Fiber in Nutrition*, Plenum Press, New York and London.
69. CONN, H. O. and LIEBERTHAL, M. M. (Eds.) (1979). *The Hepatic Coma Syndromes and Lactulose*, William and Wilkins Co., Baltimore.
70. WEBER, F. L. (1981). In: *Nutritional Pharmacology*, G. A. Spiller (Ed.), Alan R. Liss Inc., New York.
71. ADACHI, S. and PATTON, S. (1961). *J. Dairy Sci.*, **44**, 1375.
72. HOFFMAN, K., MOSSEL, D. A., KORUS, W. and VAN DE KAMER, J. H. (1964). *Klin. Wochenschr.*, **42**, 126.
73. SAHOTA, S. S., BRAMLEY, P. M. and MENZIES, I. S. (1982). *J. Gen. Microbiol.*, **128**, 319.
74. CALLOWAY, D. H., COLASITO, D. J. and MATHEWS, R. D. (1966). *Nature (London)*, **212**, 1238.
75. BIGARD, M. A., GAUCHER, P. and LASALLE, C. (1979). *Gastroenterology*, **77**, 1307.
76. KEIGHLEY, M. R. B. *et al.* (1981). *Brit. J. Surg.*, **68**, 554.
77. GALLEY, A. H. (1954). *Brit. J. Anaesth.*, **26**, 189.
78. PERTSIOUNIS, S., EGGER, G. and BIRCHER, J. (1969). In: *Symposium on Lactulose*, Baden, Austria, September 8th.
79. INGELFINGER, F. (1965). Editorial comments, In: *Year Book of Medicine*, P. B. Beeson (Ed.), Year Book Medical Publishers, Chicago.
80. BIRCHER, J., MULLER, J., GUGGENHEIM, P. and HAMMERLI, U. P. (1966). *Lancet*, **1**, 890.
81. AGOSTINI, L., DOWN, P. F., MURISON, J. and WRONG, O. M. (1972). *Gut*, **13**, 859.
82. VINCE, A. J. and BURRIDGE, S. M. (1980). *J. Med. Microbiol.*, **13**, 177.
83. BURKITT, D. P. and PAINTER, N. S. (1972). *Lancet*, **2**, 1408.
84. PAINTER, N. S., ALMEIDA, A. Z. and COLEBOURNE, K. W. (1972). *Brit. Med. J.*, **2**, 137.
85. FINDLAY, J. M., SMITH, A. N., MITCHELL, W. D., ANDERSON, A. J. B. and EASTWOOD, M. A. (1974). *Lancet*, **1**, 146.
86. HILL, M. J. *et al.* (1971). *Lancet*, **1**, 95.
87. HILL, M. J. *et al.* (1975). *Lancet*, **2**, 535.
88. EASTWOOD, M. A. and HAMILTON, D. (1968). *Biochim. Biophys. Acta*, **152**, 165.
89. BURKITT, D. P. (1970). *Lancet*, **2**, 1237.
90. POMARE, E. W., LOW-BEER, T. S. and HEATON, K. W. (1974). In: *Advances in Bile Research*, S. Mater, J. Hackenschmidt, P. Back and W. Gerok (Eds.), Schattauer Verlag, Stuttgart and New York.
91. COOKSON, F. B, ALTSCHUL, R. and FEDOROFF, S. (1967). *J. Atheroscler. Res.*, **7**, 69.
92. KEYS, A. F., GRANDE, F. and ANDERSON, J. W. (1961). *Proc. Soc. Exp. Biol. Med.*, **106**, 555.
93. JENKINS, D. J. A., LEEDS, A. R., NEWTON, C. and CUMMINGS, J. H. (1975). *Lancet*, **1**, 116.
94. KIEHM, T. G., ANDERSON, J. W. and WARD, K. (1976). *Amer. J. Clin. Nutr.*, **29**, 895.
95. MORRIS, J. N., MARR, J. W. and CLAYTON, D. G. (1977). *Brit. Med. J.*, **2**, 1307.

96. CLEAVE, T. L. (1974). *The Saccharine Disease*, J. Wright & Sons, Bristol.
97. BURKITT, D. P. and TROWELL, H. C. (1975). *Refined Carbohydrate Foods and Disease*, Academic Press, New York.
98. JENKINS, D. J. A., LEEDS, A. R., GASSUL, M. A., COCHET, B. and ALBERTI, M. M. (1977). *Annal. Int. Med.*, **36**, 20.
99. CRANE, R. K. (1960). *Physiol. Revs.*, **40**, 789.
100. FISHER, R. B. and PARSONS, D. S. (1949). *J. Physiol.*, **110**, 36.
101. WILSON, T. H. and WISEMAN, G. (1954). *J. Physiol.*, **123**, 116.
102. MILLER, D. and CRANE, R. K. (1961). *Biochim. Biophys. Acta*, **52**, 293.
103. HOPFER, U., NELSON, K., PERROTTO, J. and ISSELBACHER, K. (1973). *J. Biol. Chem.*, **248**, 25.
104. PETERS, T. J., HEATH, J. R., WANSBROUGH-JONES, M. H. and DOE, W. F. (1975). *Clin. Sci. Mol. Med.*, **48**, 259.
105. PETERS, T. J. (1981). *J. Clin. Path.*, **34**, 1.
106. STORELLI, C., VOGELI, H. and SEMENZA, G. (1972). *FEBS Letters*, **24**, 287.
107. NICHOLSON, J. T. L. and CHORNOCK, F. W. (1942). *J. Clin. Invest.*, **21**, 505.
108. MILLER, T. G. and ABBOT, W. O. (1934). *Amer. J. Med. Sci.*, **187**, 595.
109. MILLER, T. G. (1944). *Gastroenterology*, **3**, 141.
110. SPERBER, J., HYDEN, S. and EDMAN, J. J. (1953). *Annal. Roy. Agric. Coll. Sweden*, **20**, 337.
111. BORGSTRÖM, B., DAHLQVIST, A., LUNDH, G. and SJOVALL, J. (1957). *J. Clin. Invest.*, **36**, 1521.
112. FORDTRAN, J. S., CLODI, P. H., SOERGEL, K. H. and INGELFINGER, F. J. (1962). *Annal. Int. Med.*, **57**, 883.
113. JACOBSON, E. D., BONDY, D. C., BROITMAN, S. A. and FORDTRAN, J. S. (1963). *Gastroenterology*, **44**, 761.
114. HOLDSWORTH, C. D. and DAWSON, A. M. (1964). *Clin. Sci.*, **27**, 371.
115. MCMICHAEL, H. B., WEBB, J. and DAWSON, A. M. (1967). *Clin. Sci.*, **33**, 135.
116. GRAY, G. M. and INGELFINGER, F. J. (1966). *J. Clin. Invest.*, **45**, 388.
117. FORDTRAN, J. S. (1966). *Gastroenterology*, **51**, 1092.
118. BULL, J. and MENZIES, I. S. (1979). *Gut*, **20**, A449.
119. RAJAN, K. T., RAO, P. S. S., PONNUSAMY, I. and BAKER, S. J. (1961). *Brit. Med. J.*, **1**, 29.
120. HELMER, O. M. and FOUTS, P. J. (1937). *J. Clin. Invest.*, **16**, 343.
121. FOURMAN, L. P. R. (1948). *Clin. Sci.*, **6**, 289.
122. BENSON, J. A. *et al.* (1957). *New Eng. J. Med.*, **256**, 335.
123. ROBERTS, J. G., BECK, I. T., KALLOS, J. and KAHN, D. S. (1960). *Canad. Med. Assoc. J.*, **83**, 112.
124. FORDTRAN, J. S., SOERGEL, K. H. and INGELFINGER, F. J. (1962). *New Eng. J. Med.*, **267**, 274.
125. THAYSEN, E. H. and MULLERTZ, S. (1962). *Acta Med. Scand.*, **171**, 521.
126. BAYLESS, T. M., WHEBY, M. S. and SWANSON, V. C. (1968). *Amer. J. Clin. Nutr.*, **21**, 1030.
127. SLADEN, G. E. and KUMAR, P. J. (1973). *Brit. J. Med.*, **3**, 223.
128. HINDMARSH, J. T. (1976). *Clin. Biochem.*, **9**, 141.
129. STEVENS, F. M. *et al.* (1977). *J. Clin. Path.*, **30**, 76.
130. SANTINI, R., SHEEHY, T. W. and MARTINEZ-DE JESUS, J. (1961). *Gastroenterology*, **40**, 772.

131. SAMMONS, H. G. *et al.* (1967). *Gut*, **8**, 348.
≠ 132. HAENEY, M. R., CULANK, L. S., MONTGOMERY, R. D. and SAMMONS, H. G. (1978). *Gastroenterology*, **75**, 393.
133. ROLLES, C. J., KENDALL, M. J., NUTTER, S. and ANDERSON, C. M. (1973). *Lancet*, **2**, 1043.
134. WOLFISH, M. G., HILDICK-SMITH, G. J., EBBS, J. H., CONNELL, M. L. and SASS-KORTSAK, A. (1955). *Amer. J. Dis. Childh*, **90**, 609.
135. JONES, W. O. and DI SANT AGNESE, P. A. (1963). *J. Pediatr.*, **62**, 50.
136. HAWKINS, K. I. (1970). *Clinical Chemistry*, **16**, 749.
137. DUCKER, D. A., HUGHES, C. A., WARREN, I. and MCNEISH, A. S. (1980). *Gut*, **21**, 133.
138. LAMABADUSURIYA, S. P., PACKER, S. and HARRIES, J. T. (1975). *Arch. Dis. Childh.*, **50**, 34.
139. BUTS, J.-P., MORIN, C. L., ROY, C. C., WEBER, A. and BONIN, A. (1978). *J.Pediat.*, **90**, 729.
140. MONTGOMERY, R. D. *et al.* (1978). *Quart. J. Med.*, New series **XLVII**, 197.
141. ANDERSON, C. M., KERRY, K. R. and TOWNLEY, R. R. W. (1965). *Arch. Dis. Childh.*, **40**, 1.
142. MARKS, J. F., NORTON, J. B. and FORDTRAN, J. S. (1966). *J. Pediat.*, **69**, 225.
143. NOONE, C., BEACH, R. C., BULL, J. and MENZIES, I. S. (1982). *Gut*, **23**, A921.
144. MENZIES, I. S., MOUNT, J. N. and WHEELER, M. J. (1978). *Ann. Clin. Biochem.*, **15**, 65.
145. FORDTRAN, J. S., RECTOR, F. C., EWTON, M. F., SOTER, N. and KINNEY, J. (1965). *J. Clin. Invest.*, **44**, 1935.
146. FORDTRAN, J. S., RECTOR, F. C., LOCKLEAR, T. W. and EWTON, M. F. (1967). *J. Clin. Invest.*, **46**, 287.
147. LOEHRY, C. A., KINGHAM, J. and BAKER, J. (1973). *Gut*, **14**, 683.
148. MÜLLER, M., WALKER-SMITH, J., SHMERLING, D. H., CURTIUS, C. H. and PRADER, A. (1969). *Clin. Chim. Acta*, **24**, 45.
149. LAKER, M. F. and MENZIES, I. S. (1977). *J. Physiol.*, **265**, 881.
150. WHEELER, P. G., MENZIES, I. S. and CREAMER, B. (1978). *Clin. Sci. Mol. Med.*, **54**, 495.
151. COBDEN, I. *et al.* (1980). *Gut*, **21**, 293.
152. HAMILTON, I., COBDEN, I., ROTHWELL, J. and AXON, A. T. R. (1982). *Gut*, **23**, 202.
153. MENZIES, I. S. *et al.* (1979). *Lancet*, **2**, 1107.
154. BEACH, R. C., MENZIES, I. S., CLAYDEN, G. S. and SCOPES, J. W. (1982). *Arch. Dis. Childh.*, **57**, 141.
155. KENDALL, M. J., NUTTER, S. and HAWKINS, C. F. (1972). *Lancet*, **1**, 1017.
156. CHADWICK, V. S., PHILLIPS, S. F. and HOFMANN, A. F. (1977). *Gastroenterology*, **73**, 247.
157. MILLER, D. and CRANE, R. (1961). *Biochim. Biophys. Acta*, **52**, 281.
158. DAHLQVIST, A. (1962). *J. Clin. Invest.*, **41**, 463.
159. CROSBY, W. H. and KUGLER, H. W. (1957). *Amer. J. Dig. Dis.*, **2**, 236.
160. RUBIN, C. E. and DOBBINS, W. O. (1965). *Gastroenterology*, **49**, 676.
161. LITTMAN, A. and HAMMOND, J. B. (1965). *Gastroenterology*, **48**, 237.
162. PETERNEL, W. W. (1965). *Gastroenterology*, **48**, 299.
163. MCMICHAEL, H. B., WEBB, J. and DAWSON, A. M. (1965). *Lancet*, **1**, 717.

164. HAEMMERLI, U. P. *et al.* (1965). *Amer. J. Med.*, **38**, 7.
165. NEWCOMER, A. D. and McGILL, D. B. (1966). *Gastroenterology*, **50**, 340.
166. ISOKOSKI, M., JUSSILA, J. and SARNA, S. (1972). *Gastroenterology*, **62**, 28.
167. STENSTAM, T. (1946). *Acta Med. Scand.*, **125**, Suppl. 177, 1.
168. TYGSTRUP, N. and LUNDQVIST, F. (1962). *J. Lab. and Clin. Med.*, **59**, 102.
169. COZZETTO, F. J. (1964). *Amer. J. Dis. Childh.*, **107**, 605.
170. SALMON, P. R., READ, A. E. and McCARTHY, C. F. (1969). *Gut*, **10**, 685.
171. SASAKI, Y. *et al.* (1970). *J. Lab. Clin. Med.*, **76**, 824.
172. NEWMAN, A. (1974). *Gut*, **15**, 308.
173. LAWS, J. W. and NEALE, G. (1966). *Lancet*, **2**, 139.
174. LEVITT, M. D. and DONALDSON, R. M. (1970). *J. Lab. Clin. Med.*, **75**, 937.
175. CALLOWAY, D. H., MURPHY, E. L. and BAUER, D. (1969). *Amer. J. Dig. Dis.*, **14**, 811.
176. METZ, G., JENKINS, D. J. A., PETERS, T. J., NEWMAN, A. and BLENDIS, L. M. (1975). *Lancet*, **1**, 1155.
177. METZ, G., JENKINS, D. J. A., NEWMAN, A. and BLENDIS, L. A. (1976). *Lancet*, **1**, 119.
178. COOK, G. C. and HOWELLS, G. R. (1968). *Amer. J. Dig. Dis.*, **13**, 634.
179. MENZIES, I. S. and SEAKINS, J. W. T. (1976). In: *Chromatographic and Electrophoretic Techniques*, Vol. 1, Ivor Smith and J. W. T. Seakins (Eds.), William Heinemann, London.
180. MENZIES, I. S. (1973). *J. Chromatog.*, **81**, 109.
181. LAKER, M. F. (1979). *J. Chromatog.*, **163**, 9.
182. BOND, J. H. and LEVITT, M. D. (1975). *J. Lab. Clin. Med.*, **85**, 546.
183. SALMON, P. R., AJDUKIEWICZ, A. B., CLAMP, J. R. and READ, A. E. (1970). *Gut*, **11**, 1065.
184. KING, C. E., TOSKES, P. P., SPIVEY, J. C., LORENZ, E. and WELKOS, S. (1979). *Gastroenterology*, **77**, 75.
185. Leader: Sugaring the Crosby Capsule (1981). *Lancet*, **1**, 593.

*Chapter 5*

# RECENT DEVELOPMENTS IN NON-NUTRITIVE SWEETENERS

J. D. HIGGINBOTHAM

*Tate & Lyle Group Research and Development, Reading, UK*

*SUMMARY*

*There can be few areas of new product development that arouse more interest or controversy and which receive more industrial effort than high-intensity sweeteners. This chapter seeks to illustrate this by describing recent progress and current achievements in two groups of sweeteners. One group includes the general-purpose, wide-utility sweeteners and the other group contains the more specialised sweeteners with other properties besides sweetness. Selection for inclusion in both groups is restricted to those sweeteners which are in commercial development or are close to being marketed. The emphasis is on their practical applications reflecting their individual advantages or shortcomings. Some comparisons are made between the sweeteners to draw attention to their relative importance or range of uses and their competitive market potential.*

## 1. INTRODUCTION

The years from 1979 to 1983 have seen the culmination of many years of development of three non-nutritive high-intensity sweeteners, namely aspartame, Acesulfame-K®* and thaumatin or Talin® Protein Sweetener† (Talin). These are now permitted for use in foods in several countries after

---

* Acesulfame K® is a registered trade mark of Hoechst AG, West Germany.
† Talin® Protein Sweetener is a registered trade mark of Tate & Lyle PLC, UK.

up to 14 years of safety evaluation, although it will be several more years before their regulatory approval is widespread. They provide acceptable alternatives to the two best-known sweeteners, saccharin and cyclamate, although these are still not bettered, or even matched, for cost, stability and breadth of application. The controversial and long-running issue of the safety of saccharin and cyclamate has still not been fully resolved, but restricted use in many countries is permitted. There is sufficient concern about their safety, however, that if a particularly good, safe and practical alternative sweetener acceptable to the food and drink industry were to be developed then the need for their continued use would fade and lead to their withdrawal in most countries. This situation is far off, and for the foreseeable future industry and the consumer will have a range of compromises to choose from, each sweetener or combination of sweeteners finding specialised applications which best suit its particular properties.

This chapter does not claim to be comprehensive or exhaustive as a catalogue of interesting or novel, strongly sweet-tasting substances. The intention is to describe the developments that have occurred in the fast-moving field of sweeteners since the previous publication in this series. Little emphasis will be placed on structure–sweetness relationships or the properties of synthetic analogues or derivatives of the major sweetener groups. The research literature is vast in this area and has been well reviewed.[1] The main emphasis will centre on those sweeteners which have actual or potential commercial applications and which provide practical alternatives either alone or in combination. Some of their practical uses will be described and their individual competitiveness, availability and future potential will be evaluated.

The discussion of individual sweeteners follows in approximate order of their relative importance. They fall into two main groups: the first, which could be termed the general-purpose high-intensity sweeteners, includes saccharin and cyclamates, and the more recently introduced aspartame, Acesulfame-K, chlorosucroses and stevioside; the second, a more specialised group of sweeteners which have a poorer sweet-taste quality than the first group but which have other potentially redeeming properties such as flavour enhancement and/or natural origin, includes Talin, the dihydrochalcones and glycyrrhizin. Finally, a brief mention will be made of some interesting new sweeteners.

## 2.   SACCHARIN

Saccharin has been used as a sweetener for over 80 years. It is estimated that, despite all the adverse publicity about its potential hazards to health, some 70 million Americans are fairly regular users, including a third of children under the age of 10. About 60 % is consumed in soft drinks, 20 % in other beverages and foods, and 20 % as table-top sweeteners. The Food and Drug Administration (FDA) estimates that nearly 3000 tons of saccharin (out of around 5000 tons worldwide) is consumed in the USA. When it is realised that the sweetness of saccharin is about 300 times that of sucrose then almost 1 million tons of sugar are being substituted. The saccharin market (that is, products containing saccharin) is valued at £1000 million annually in the US alone and is still increasing.

### 2.1. Saccharin as a Sweetener

The success of saccharin can be attributed to several factors: first, it is cheap—about 1/20th the price of sugar in sweetness-equivalent terms— and simple to produce; second, it is not metabolised and provides no calories; third, it does not affect the teeth; and fourth, it is stable in use and therefore has a wide range of applications.

Despite these advantages, saccharin still has several drawbacks. Since cyclamate was banned in the USA in 1969, the mixture of saccharin and cyclamate is no longer available, but saccharin alone has undesirable metallic or bitter aftertastes. People continue to use it, many being unable to detect it when mixed with sugar, while some even prefer it to sugar, especially in hot beverages. However, many patents exist which claim reduction in aftertaste, suggesting that the sweetness quality needs improving, particularly in low-calorie saccharin-only soft drinks where the aftertaste is more noticeable.

### 2.2. The Safety Question

Ever since the Canadian two-generation rat studies in 1977 indicated that saccharin was carcinogenic to rats when consumed at levels equivalent to an average-weight human drinking 1250 cans of diet soft drinks per day (see Reference 53, Chapter 6), there has been an unresolved question about its safety. These results, following two earlier incriminating studies, prompted Canada to ban saccharin for most uses and in April 1977 the FDA announced its intention to revoke the approval of saccharin in foods, except for diabetics and the obese if it could be shown to be strictly necessary. This evoked a public outcry, the FDA alone receiving 100 000 public comments—mostly opposing the ban—and Congress imposed an 18-month moratorium on the ban becoming effective, ostensibly to allow more definitive studies to be completed. This moratorium has had two extensions, the most recent lasting until August 1983.

Meanwhile, many users, manufacturers, and associations such as the

Calorie Control Council, have mounted strong campaigns to support saccharin and other low-calorie alternatives, assembling a formidable bank of scientific and epidemiological data in their support and funding several safety studies. Their most notable study (not yet published) is the largest chronic study in the history of saccharin, using thousands of rats over two generations. This and two other key studies will be completed by late 1982 and should be reported in 1983. These studies may raise more questions than they resolve as it seems that toxicologists and regulatory bodies find agreement difficult.

The UK report on sweeteners[2] recommended continued restricted use of saccharin and in Europe, the Joint Expert Committee on Food Additives (JECFA) is due to review its temporary acceptable daily intake (ADI) but may delay until March 1984 to have more data from the ongoing studies in the USA.

### 2.3. Some Consumer Saccharin Products

For consumer use, saccharin is available as a bulked solid mainly for restaurant and institutional purposes, such as the Sweet 'N Low®* sachet formulation which has lactose and cream of tartar to depress the aftertaste. The small tablets delivered singly from plastic dispensers are popular for beverage sweetening but more recently liquid products, specially formulated to reduce aftertaste, have been heavily promoted and have been found to taste better than saccharin alone. For cooking and easy sprinkling, various dry mixtures with sucrose are marketed, often with reduced bulk density to give the equivalent sugar-sweetness level per spoonful. In Spain, Linea®,† a mixture of saccharin, fructose and mannitol formulated to give three times sucrose sweetness, is upsetting the sugar industry by being marketed as 'the sugar without sugar!'

### 2.4. Future Prospects for Saccharin

It seems that, except in the unlikely event of a complete ban, saccharin use will continue to flourish and achieve a balance with the other sweeteners as they are introduced. It will take some knocks from companies who decide to remove it from their products for marketing, image or labelling reasons, from individual countries such as Nigeria which intends to ban it in 1983 and Denmark and Holland which wish to restrict its intake, and from a

---

* Sweet 'N Low® is a registered trade mark of Cumberland Packing Corp., USA.
† Linea® is a registered trade mark of Laboratories Wasserman, Spain.

growing number of people who decide to reduce their intake of synthetic food additives.

Sherwin Williams, the world's largest, and the only US, producer is sufficiently confident to increase its production capacity in Cincinnati and is developing calcium saccharin as an alternative to the sodium salt as a consequence of growing concern about dietary sodium intake.

## ✓ 3.  CYCLAMATES

Cyclamate, as the sodium or calcium salt, rose rapidly in popularity during the 1950s and 1960s to become the dominant artificial sweetener, despite its higher unit cost compared to saccharin. Although only around 30 times sweeter than sugar, it has the particular advantage of being sweeter when mixed with saccharin and of reducing the overall saccharin aftertaste, giving a widely acceptable sweet taste. Over 9000 tons were consumed annually in the USA before it was banned in 1970 based on bladder tumours found[121] in saccharin:cyclamate combination two-year feeding studies reported by Abbott Laboratories to the FDA in October 1969. The UK and Japan followed suit with a total ban, while other countries restricted its use although consumption in Europe remains at up to 2000 tons per annum. Clearly this is another example of widely diverging opinions in toxicology interpretation.

### 3.1.  The Argument for the Greater Use of Cyclamate
For 10 years after the US ban, Abbott Laboratories pressed strongly for the reintroduction of cyclamate, until the FDA rejected a petition for its approval in September 1980. Interestingly, the reason given was not that cyclamate or its occasional metabolite cyclohexylamine were proven carcinogens or mutagens but rather that the data had not demonstrated adequately that cyclamate was totally safe. Abbott decided not to appeal or continue the fight but to wait until the regulatory climate improves and pressure develops following gradual reinstatement in other countries.

At their meeting in May 1982, JECFA effectively raised the ADI of cyclamate by allowing a 50-fold safety factor (rather than the previous 100-fold) and removed the temporary status to give a full ADI of $0-11\,mg\,kg^{-1}$. This offers some encouragement to cyclamate producers and users but still does not accommodate widespread use at this level and will have little influence in the UK and USA.

The recent UK sweeteners review[2] recommended that cyclamates should remain prohibited because of the still unknown effects of cyclohexylamine in man and because its relatively low sweetness would result in a high intake, especially in children, if it were to be reintroduced into soft drinks. Also the availability of Acesulfame-K and aspartame suggested that there was less need for cyclamate. For the present, the cyclamate issue is dormant but showing signs of life. With the Calorie Control Council, the American Diabetes Association and some sectors of industry still supporting its reintroduction, cyclamate may one day again be an active participant in the sweetener business.

## 4. ASPARTAME

In July 1981 the FDA finally announced limited approval of aspartame, which took effect in October of that year. This represented a triumph for G. D. Searle, the manufacturers, who had fought a continuous battle since 1974 when the initial regulatory approval was challenged, preventing its being marketed for seven years while arguments over safety and even scientific honesty were debated. The Canadian authorities approved its use in August 1981 in soft drinks in addition to the US categories of permitted use (table-top sachets, tablets for hot beverages, instant coffee and tea, chewing gum, powdered drinks, cold cereals, gelatins, puddings and dessert toppings). This major application for sweeteners was denied in the USA because of aspartame's main drawback, its instability to the low pH (especially below pH 3) of soft drinks. Aspartame is a dipeptide ester of aspartic acid and phenylalanine which can hydrolyse and cyclise in aqueous conditions especially below pH 3 or above pH 6 to form the corresponding diketopiperazine which is not sweet. Searle states that in practice there is no noticeable loss of sweetness when formulated at a slightly higher pH, even over five months, which gives sufficient shelf-life for soft drinks. The company expects to re-petition the FDA to also allow the use of aspartame in soft drinks in the USA but there are technical problems to solve first and there is some concern about possible interactions of its breakdown products with food and drink ingredients.

Several aspartame-sweetened drinks have been introduced to the Canadian market with vigorous advertising and are generally well received, especially by diabetics and serious dieters who previously had no acceptable products available. There are some suggestions that the products are too sweet—possibly intentional to allow for partial loss of sweetness on storage before consumption—and that the quality of taste, although good, is somewhat 'sickly' sweet after drinking a normal quantity. The table-top sachet marketed as Equal[®]* by Searle, a lactose/aspartame/silica mixture, has a good quality in hot beverages. Other products, such as powdered beverages, a lemonade mix and a low-calorie lemon-tea made by Lipton's are now being test marketed in the USA.

## 4.1. Aspartame Manufacture

### 4.1.1. Chemical Processes

In 1965, James Schlatter of G. D. Searle was credited with being the first to notice the intense sweetness of the dipeptide ester. Soon afterwards (April 1967) Searle filed its master patent[3] on the use of aspartame as a sweetener, the priority of which expires in 1983 in the UK. An extension of the patent life may be granted to compensate for the long period (seven years) of regulatory delay.

However, Searle was unable to claim aspartame itself as a new material because this dipeptide ester was already a known compound[4] prepared in the ICI Laboratories in 1965 as an intermediate in gastric tetrapeptide synthesis. The process described was not suitable for commercial production of aspartame, and neither were classical peptide-synthesis techniques using protected amino acids, activating groups and subsequent de-protection. In 1970 Ajinomoto Co. of Japan described a more practical process[5] which was subsequently improved.[6] This invention did not require protecting groups or expensive reagents and gave reasonable overall yields. Both $\alpha$- and $\beta$-dipeptides are formed in an 80 % yield in the ratio of 6:4. The desired $\alpha$-aspartic acid derivative was induced to crystallise preferentially to give 33 % yield overall. Other examples in this patent yield up to 80 % of the $\alpha$-isomer using mixed organic solvents at low temperature. The cheapest and simplest chemical route is probably achieved by the reaction of L-aspartic anhydride hydrochloride with L-phenylalanine methylester in acid conditions in organic solvents at low temperature. Since this route was described Ajinomoto has published some 40 patents describing modifications and improvements, introducing novel methods of coupling, protecting and purification while minimising diketopiperazine formation. Other companies, notably Monsanto,[7,8] have filed patents on competing methods for aspartame synthesis using $N$-protected L-aspartic acid anhydrides with L-$\beta$-phenylalanine in glacial acetic acid. Stamicarbon[9] describes the use of aqueous solvents, also previously published by Ajinomoto[6] in their 1971 patent. Beecham filed a German patent[10] (since

* Equal[®] is a registered trade mark of G. D. Searle & Co., USA.

abandoned) in 1970 describing a similar process but using esters such as ethyl acetate as a solvent.

### 4.1.2. Enzymic Processes

Now that aspartame is a commercial reality, the incentive to improve and reduce the cost of synthesis has led to a recent announcement by Toyo Soda[11] suggesting the use of enzymes to reduce the cost within one or two years. Toyo Soda's original patent[12] was filed in 1978 suggesting that development has been slow and is still at the pilot stage. The enzymic process uses a microbial protease, such as thermolysin, to link $N$-protected aspartic acid to $\beta$-phenylalanine methyl ester. The specificity of an enzyme offers considerable advantages over chemical processes since racemic mixtures of the amino acids can be used with only the L-isomer reacting.

The process also forms insoluble addition compounds which force the enzymic reaction to go to completion and provide an excellent purification step. Recovery of protected aspartame is easily achieved, giving high overall yields and good purity. Searle is also investigating enzyme processes[13] in which proteolytic enzymes with esterase activity selectively alkylate the dipeptide.

### 4.1.3. Genetic Engineering Processes

A dipeptide ester such as aspartame should be amenable to synthesis using genetic engineering techniques. The Cetus Corporation describes[14] a somewhat theoretical approach in which an alternating polypeptide of phenylalanine and aspartic acid is produced which is then hydrolysed and methylated chemically to aspartame. The main advantage appears to be that inexpensive, racemic mixtures of raw materials can be used rather than the pure L-amino acids. A similar route has been described by Searle.[15] Although promising, genetic engineering methods still need to prove themselves as commercially viable, and may require further toxicological evaluation.

### 4.2. The Development of Aspartame in Foods, Drinks and Oral-Care Compositions

#### 4.2.1. Improvement in Handling Dry Products

In the early 1970s Searle offered aspartame to General Foods to help develop applications and overcome some of the practical difficulties of using this sweetener. This resulted in a number of patents being assigned to General Foods, many of which were designed to improve the dry-storage and flow characteristics of the dry crystalline powder and to improve the rate of dissolution in cold water. Aspartame tends to separate from dry mixes. Blending in a slurry with monocalcium phosphate helps alleviate this problem[16] and fusing in a matrix of citric acid,[17] for example, allows dry-beverage formulation by improving solubility and reducing dry-ingredient interactions with some flavours. Dry grinding together with a food acid such as citric or malic acid improves the rate of solution,[18] as does co-drying with a bland, low-calorie polysaccharide such as polymerised dextrin.[19] Other suggested solutions to the problem include using an edible gum, such as carboxymethylcellulose, and crystalline sugar with cola flavour and monocalcium phosphate in a matrix to give a rapidly soluble diet-beverage composition.[20] Co-drying 0·5–10 % aspartame with a starch hydrolysate at DE ≤ 20, especially sorghum starch,[21] or aggregating aspartame crystals with a small amount of water to produce a flowable, dry sweetening composition which is also rapidly soluble, is claimed as patentable by General Foods.[22]

### 4.2.2. Sweetener Mixtures

Numerous patents describe mixtures of aspartame with other high-intensity sweeteners, notably saccharin and cyclamate, to take advantage of taste quality improvement, synergy of sweetness and reduction of unit cost. For example, a 4:1[23] or 6:1[24] blend of aspartame and saccharin is claimed to be 1000 times sweeter than sucrose. This presumably is a threshold sweetness value since Searle[25] quotes a 5:1 ratio as some 300 times sweeter than sucrose at 10 % concentration. A purely additive combination would provide just over 200 times the sweetness of sucrose, so considerable synergy is evident with significant saving in cost.

### 4.2.3. Liquid Formulations and Dental Uses

Liquid applications require special techniques to improve the stability of aspartame and minimise the formation of breakdown products with loss of sweet taste. Liquid or semi-liquid applications would include oral-care products, e.g. mouthwashes and toothpastes, chewing gum, and liquid-sweetener concentrates. Unlike soft drinks where the pH and the aspartame concentration are low, an optimum pH for stability (around pH 4) can be selected for these products, and compounds reducing hydrolytic and bacterial activity in water can be included. Propylene glycol can be used to stabilise liquid concentrates,[26] or mixtures of up to 5 % aspartame in ethanol, glycerol, propylene glycol or butylene glycol[27] can be made for direct addition of a liquid sweetener to hot tea or coffee.

Formulating an anhydrous dentifrice with aspartame, an anti-caries agent and liquid vehicles entrapped in an acid anhydride stabiliser at pH 3–5 is claimed to maintain stability.[28] Aspartame may be kept separate in a gum with ethylene glycol and mixed with encapsulated aqueous and base components only when the teeth are brushed.[29]

Besides adding sweetness and palatability to dentifrices, aspartame is claimed[30] to be able to reduce the incidence of dental caries by reducing tooth demineralisation. Citing examples from a wide range of food and drinks, it is claimed that the addition of aspartame reduces tooth-enamel demineralisation and caries severity not only by replacing or reducing sucrose levels but also when used in the presence of normally cariogenic carbohydrates. As an anti-cariogenic agent, the incidental addition of the extra sweetness of aspartame can be reduced by the addition of alum or other sour-tasting substances. The mechanism of this caries-inhibiting effect of aspartame may be a buffering effect on the acids formed by oral bacteria or simply a stimulation of salivary flow, liberating calcium and raising oral pH levels.

### 4.2.4. Chewing Gum and Flavour Enhancement

To return to specific applications where the stability of aspartame can be enhanced, chewing gum can be formulated at pH 5–7 to minimise decomposition[31] or the aspartame can be kept separate by encapsulating it in polysaccharide[32] which gives longer-lasting sweetness and flavour[33] —however, some flavours are not compatible because of adverse interaction with aspartame's ester group. Aspartame is found to interact positively with certain fruit flavours, such as lemon and strawberry, to intensify their perception, particularly in drinks and gelatin desserts.[34] Also, addition of aspartame below the sweetness threshold concentration (0·001–0·02%), enhances fruit flavours and aromas.[35] Baldwin and Korschger[36] showed that both orange- and cherry-flavoured beverages had intensified flavours although no enhancement was noted in strawberry-flavoured drinks. However, no differences were noted in these flavours in gelatin desserts although the sweetness of aspartame was significantly greater in gelatin (214 times at 15% sucrose equivalent concentration) than in the beverages (146 times at 9·5% sucrose equivalent concentration). Searle's own taste-panels found[25] a potency of 160 times at 18% sucrose equivalent concentration in gelatin, also suggesting greater than expected sweetness at such a high sucrose level. Sweetness perception is particularly high in iced milk desserts—some 280 times that of sucrose at 14% sucrose equivalent concentration, attributed to synergy with sorbitol.

*4.2.5. Table-Top Sachets and Tablets*

Table-top sweeteners are ideal applications for aspartame when dispensed from sachets because the product is protected from hygroscopicity by the sachet and by admixture with lactose and silica. Some patents issued in the early 1970s suggested[37] that the slightly flat sweetness and hint of bitter aftertaste could be improved by the addition of potassium bitartrate plus lactose or dextrose. A non-caloric alternative was also proposed[38] at the same time, using glucono-$\delta$-lactone and sodium gluconate to replace the carbohydrates. Another more recent formulation[39] uses glycyrrhizin to reduce the cost of aspartame by an apparent synergy which doubles the sweetness perception of the mixture. More rapid dissolution of aspartame in hot beverages without leaving foam or scum is achieved by blending mixtures of bulking agents and de-foaming compounds.[40] Ajinomoto, in Japan, add sodium glutamate or nucleotides which are said to improve the aftertaste of aspartame.[41]

Other permitted uses of aspartame are for cold breakfast cereals, for which a coating of pyrodextrin,[42] dextrin or starch hydrolysate containing 2–36 % aspartame is suggested.[43] Another use of aspartame is for improving the taste of low-grade coffees by masking bitterness.[44] A Japanese patent[45] describes the use of aspartame for a sweetened substitute for adding to hot beverages which also improves their flavour.

A tablet sweetener called Canderel®* has been on sale in France since 1979—the first permitted use of an aspartame product—while Belgium and Luxemburg followed suit in early 1980. Each tablet contains 20 mg of aspartame (equivalent to 3 g sucrose) with L-leucine, lactose and carboxymethylcellulose as fillers, and fumaric acid and sodium bicarbonate to render the tablet effervescent to aid dissolution. At the price quoted in 1980, this is the equivalent of around 10 times the price of sucrose. In 1980, Canderel sales in France were the equivalent of some 330 tons of sucrose compared with over 14 500 tons of sucrose equivalent in saccharin sales. However, in 1981 the market for Canderel was estimated to have increased six-fold, with a 6 % drop in saccharin sales, suggesting that Canderel is replacing saccharin as well as sucrose.[46]

*4.2.6. The Regulatory Position*

Many countries have now given approval to aspartame for limited table-top use following confirmation of the generous ADI of 40 mg kg$^{-1}$ recommended by JECFA,[122] including Brazil (which also allows the use of

* Canderel® is a registered trade mark of Laboratoires Searle, France.

aspartame in soft drinks), Mexico, the Philippines, Switzerland, France, Belgium, Luxemburg, Algeria, Tunisia and most recently Australia, Norway, Singapore, South Africa and West Germany. The UK review[2] also recommended food-additive uses but excluded soft drinks. The USA and Canada are insisting on 'post-marketing surveillance' to monitor how much aspartame is being sold and in which products, and to see if there are any particularly high consumption groups.

*4.2.7. Availability and Market for Aspartame*

In 1982 demand for aspartame was being met from stocks prepared before the originally anticipated launch in 1975, while production plants in Tokyo and South Chicago were being restarted with an anticipated production capacity of 230 tons in 1982 and 1800 tons by 1983.[47] At present prices of around US$200 per kilogram for large purchases, if all production were to be sold, sales valued at US$46 million rising to US$360 million would be achieved. *Business Week* went so far as to suggest that sales of US$1700 million could be achieved by displacing *all* saccharin and some of the sugar market in soft drinks.[48] This would seem most unlikely as the *circa* 9000 tons of aspartame this represents is equivalent to some 4500 tons of saccharin in sweetening power, more than twice the present US soft-drink consumption of saccharin. The total present value of saccharin sales in the USA is about US$30 million per annum, so to achieve sales of 9000 tons of aspartame twice the weight of saccharin currently used at 22 times its price must be consumed. Other imponderables are the effect of warning labels for phenylketonurics, and the still unresolved problem of stability in soft drinks.

A recent estimate of the aspartame market in 1990 is given by Genex in Rockville, Maryland, who are quoted[49] as stating that some 30 million lbs per annum of aspartic acid (of which they are manufacturers) will be required. This is equivalent to some 5500 tons of aspartame based on a 49–50 % overall yield from aspartic acid.

However, the better taste and image of aspartame, future cost reductions and possible restrictions on saccharin use should ensure that substantial sales begin to repay the vast costs of the development of aspartame.

## 5.  ACESULFAME-K

Next in this group of sweeteners is the relative newcomer, Acesulfame-K. Discovered by Hoechst[50] in 1967, this acetoacetic-acid derivative, similar in structure to saccharin but half as sweet at 150 times the sweetness of sugar at 3–4% equivalent concentration, has been developed steadily with little publicity. It is clear that it was viewed seriously as a sweetener for during the last 14 years some 40 toxicological studies have been completed,[2] including three long-term chronic studies (mouse, rat and dog) in the Netherlands, and others in Germany, the USA and Switzerland. Acesulfame-K is apparently not metabolised.[51] Recently a special diabetic rat study has been undertaken with no adverse effects. Acesulfame-K appears to be stable in the pH range of foodstuffs under normal storage conditions. Below pH 3, however, after prolonged storage, especially at elevated temperatures, some decomposition may be noted. The main hydrolysis product, acetoacetamide, is also being checked for safety.[52] Short-term studies showed no cause for concern over this product nor over several other minor acid-degradation products.

## 5.1. Practical Applications for Acesulfame-K
### 5.1.1. Oral Hygiene and Medicines
Since 1979, all the data required for a drug master-file in the USA have been available for use in cosmetics and pharmaceuticals.

The use of Acesulfame-K in toothpastes is clearly justified as its structure indicates it to be non-cariogenic. Indeed, a recent study by the University of Kentucky College of Dentistry,[53] suggests that Acesulfame-K actually inhibits dental caries by interfering with the growth of *Streptococcus mutans* and restricting its ability to ferment dietary carbohydrates. No levels of use were reported, and the observation that saccharin and sugar alcohols also had this inhibitory effect leads to the conclusion that this is mainly of academic interest and would have no real effect on the complex population of organisms that exists in the mouth, when used at the very low levels needed in practice.

Hoechst[54] has described the utility of Acesulfame-K in toothpastes mouth sprays, mouthwashes and chewing gum. A level of 0·2% was recommended as highly desirable in toothpaste by taste panellists as it masked the taste of the surfactant used. Storage stability was excellent with no problems of solubility or flavour interference. In mouth sprays and mouthwashes the sweetness was 'fuller, richer and more balanced' than saccharin at equivalent sweetness levels.

### 5.1.2. Food and Drinks
Hoechst has concentrated on safety data accumulation, perfecting the chemical synthesis route and optimising costs and production capacity—

believed to be several hundred tons per annum. Hoechst has less experience in product applications and formulations for the food industry and the time is now ripe for cooperative developments in this area. The stability of Acesulfame-K to cooking temperatures (a distinct advantage over several other contenders) and the availability of low-calorie bulking agents, such as polydextrose, suggest that useful slimming-aid products can be made.

When Acesulfame-K is used as the sole sweetener in highly sweetened drinks such as cola at 8–10% sucrose equivalent concentration then its sweetness factor drops to about 100 times that of sucrose. This reduces its cost-effectiveness in such drinks but a reported[55] sweetness synergy with cyclamate could help to redress this if such combinations were to be permitted.

Our taste-panel assessment of soft drinks sweetened with Acesulfame-K showed a slight (though not statistically significant) preference over saccharin in lemonade, cola and tonic water. Some bitter and astringent aftertastes were noted, especially in tonic water. To give the equivalent sweetness to normal sucrose levels, the apparent sweetness intensity of Acesulfame-K was 170, 110 and 140 times that of sucrose, respectively. In tea, panellists who normally take saccharin showed a preference for Acesulfame-K-sweetened drinks although some noted an increase in astringency.

### 5.1.3.  The Regulatory Position

Hoechst filed a FDA food-additive petition in the summer of 1982 covering several food applications. Despite its apparent safety, in 1981 JECFA[122] raised some questions concerning data from chronic studies on Acesulfame-K and deferred a decision to recommend an ADI. Only specifications were being considered in 1982 and ADI considerations will await the 1983 meeting. A DNA-binding study has been completed by Hoechst[52] to help clarify the data interpretation.

Food-additive petitions concerning Acesulfame-K are lodged with all the major industrialised nations, including Canada, West Germany, France, The Netherlands, Switzerland, Belgium, Denmark, South Africa and Australia. The USSR has allowed its use in toothpaste. A German decision on its use in low-calorie dietary products is expected soon, and in France long-term clinical trials have commenced. The UK has been the first country to recommend widespread use in foods,[2] including soft drinks; a bold decision in view of the JECFA reservations and a reflection of the careful and comprehensive safety data submission. Some 34 documents were submitted to the Committee on Toxicology, dated from 1973 to 1980.

What is the potential of Acesulfame-K in competition with the other major sweeteners? It has the advantage of a slightly better taste than saccharin, having a less unpleasant aftertaste, but it is considered somewhat inferior to aspartame in taste quality and 'image'. However, it has better aqueous and heat stabilities than aspartame in most products and at a suggested price of £12–15 per kilogram it is approaching a quarter of the price of sugar, considerably less than aspartame, but is some way from competing with saccharin in price. When the food industry was fearing a possible saccharin ban in the late 1970s, Acesulfame-K received strong support, confirmed by the FACC report.[2] Now, with less pressure on saccharin, it has to compete directly with this cheaper, long-established product.

## 6. STEVIOSIDE

Stevioside, relatively unknown in the West, is widely used as a high-intensity sweetener in Japan where consumption has more than doubled since 1979—some 150 tons having been consumed in 1981. This is equivalent to over 22 000 tons of sucrose and, at a price of about £25 per kilogram, the market value of stevioside is around £3·5 million per annum. This rapid rise in consumption is due to its low price relative to sugar (about 29 % of the price of sugar in Japan), and its natural provenance—it is extracted from the leaves of *Stevia rebaudiana* which is native to Paraguay but grown in Japan and Korea with greatly improved horticulture and production methods. New varieties and hybrids are grown in southern Japan which yield twice as much as native plants and contain greater proportions of the preferred stevioside anomers—the rebaudiosides. In fact, stevioside itself is rarely used as it has low solubility (0·12 % in water at room temperature) and an unpleasant chemical/liquorice aftertaste at higher concentrations. A typical commercial 'stevioside' contains 55 % rebaudiosides and 45 % stevioside.

### 6.1. Preparation of Stevioside

Very many Japanese patents exist describing processes for the extraction of the leaves of *S. rebaudiana*, and for the subsequent separation of stevioside and the rebaudiosides from the crude extract and their subsequent crystallisation and drying. As a first step,[56] dried stevia leaves are extracted with eight parts of hot water at 50 °C, giving a black, evil-smelling, bitter crude extract. This is partially purified by precipitation with organic solvents such as ether or organochlorine compounds[57] or, more commonly, by selective removal of impurities by absorption on ion-exchange resins.[58] Further purification can be achieved by crystallisation of the product from aqueous methanol where differences in solubility between stevioside and the rebaudiosides can be exploited.[59] Stevioside crystallises selectively from 40–90 % aqueous methanol while rebaudioside A separates preferentially from the mixture in 80 % aqueous ethanol.[60] Such techniques allow the formulation of the optimum mixture of glycosides to give the best sweetness intensity, quality and solubility at an economic price.

Totally synthetic routes to stevioside and rebaudiosides A, D and E are now known,[61,62] and new derivatives of stevioside which do not degrade to the physiologically active aglycone steviol were investigated by Dynapol as potentially useful, safe new sweeteners.[63] Cell-culture techniques are being developed and callus cultures have been produced.[64] Plant breeding has increased stevioside concentrations from around 6 to over 12 % of leaf weight and improved the ratio of rebaudioside A.[65] The effect of post-harvest changes on these ratios has also been studied.[66]

## 6.2. The Taste of Stevioside

Disagreement exists over the sweetness intensity and taste quality of stevioside. Much of the confusion can be attributed to differences in composition between the commercial 'steviosides' which vary from dried crude extracts to pure crystalline glycosides. Also, in common with other intense sweeteners, its relative intensity depends on concentration as well as other variables such as temperature, pH and the presence of other ingredients. The sweetness of 'pure' stevioside (about 90 %) is quoted as about 300 times that of sugar but trained taste-panellists suggest that stevioside is 240 times sweeter than sucrose at near threshold levels (2 % sucrose), reducing to 128 times relative to 6 % sucrose.[67] The sweetness is accompanied by slight bitterness, some astringency and aftertaste with low acceptability compared to sucrose. These results are in general agreement with those of our own laboratory and of Dynapol.[68] The rebaudioside A glycoside is 20–30 % sweeter than stevioside and has better stability and a more sucrose-like taste. The other glycosides, known as dulcosides A and B, have a much lower sweetness, of the order of 36 times that of sucrose at 2 % concentration.[67] Pure rebaudioside A is sold in Japan at 30 000 yen (£73) per kilogram as 300 times sweeter than sucrose. This makes it one-and-a-half times as expensive as the stevioside/rebaudioside 44/55 % mixture which costs £24 per kilogram and has a sweetness of 150 times that of sugar. This commercial grade of 'stevioside' is thus less than one-third the price of Japanese sucrose (at 230 yen per kilogram) which explains the rapid increase in its use in new product applications.

Despite mixing the glycosides in what seems to be every imaginable combination, stevioside still has a less than optimal taste. Many patents describe attempts to improve this by admixture with histidine to mask bitterness,[69] with hydrocolloids,[70] with other high-intensity sweeteners such as thaumatin,[71] or even with sucrose itself.[72]

### 6.3. The Use of Stevioside in Japan

Stevioside is used extensively in chewing gum, both as the purified material and as ground dried stevia leaves, such as in Kanebo's Play Gum®.* It can be used in tobacco on its own as a crude stevia extract or together with other plant extracts, such as tamarind and euphorbia, or free amino acids.[73,74] The sugars and amino acids in these plant extracts, when heated with polyhydric alcohols, combine to improve the overall tobacco flavour.

Japanese cuisine uses many highly salted dishes in which stevioside has found widespread use to cut the salt taste without providing colour. Sugar in high-salt and low-pH conditions in the presence of vegetable products, e.g. as in pickles, can discolour on storage through Maillard reactions. Saccharin is occasionally used but now stevioside, in combination with sugar or glycyrrhizin, is more often used in pickles, soy sauce, and mashed and steamed fish such as kamaboko and fish sausage. Unlike sugar, of course, it does not ferment so the delicate taste of some Japanese fermented products does not alter when stevioside is used.

More familiar sweetener uses of stevioside are in soft drinks, where its relatively low price allows effective savings on sugar use, but a maximum of only 50% substitution is possible before too obvious a taste difference becomes evident. Pepsi Cola in Japan uses it with 50% fructose in diabetic drinks, but admits it does have a poor taste. Coca Cola has switched to α-G-Sweet, an enzymically modified stevioside[75] (made by Toyo Sugar Refining Co.) which is said to taste better although it is half as sweet and more expensive in use. Little, if any, toxicology testing has been undertaken.

The stability of stevioside at the low pH of soft drinks is adequate and its use in low-calorie soft drinks is likely to increase, especially if another high-intensity sweetener such as aspartame can be permitted for use in combination with it.

A very wide variety of liquid and solid powder or granule table-top

---

* Play Gum® is a registered trade mark of Kanebo KK, Japan.

sweeteners is available, based on mixtures of sugar alcohols or sugars and intensified with saccharin, glycyrrhizin and stevioside. My Home Sugar Strong®,* for example, is a stevioside, sugar and 'natural product' mixture packed in sachets of 2 g, equivalent to 10 g of sucrose, which retails at 3–5 times the normal sugar price. Sweetener tablets can be made by compressing 94 parts of fructose with 1 part of stevioside and 5 parts of a $\beta$-cyclodextrin to form a tightly packed but water-soluble moulding.[76]

Stevioside has not yet found significant applications in pharmaceuticals or oral-care products because the Japanese drug regulations usually require more stringent testing for these than for food products.

### 6.4. Future Developments of Stevioside

Interest in stevioside is increasing in the West, and it is reported[77] that Atomergic, a specialist chemical company in Long Island, New York, has petitioned the FDA for its approval. The University of Illinois Medical Center is working with the National Institute of Dental Research (NIDR) to study potential uses of stevioside in the USA. The NIDR showed stevioside to be non-cariogenic several years ago but the FDA require the usual stringent short- and long-term safety data on stevioside and the other glycosides before considering its use. It is not included in the present UK review of sweeteners[2] through lack of industrial support or toxicological data and it is not on the present agenda for JECFA review. However, the Japanese developments will focus more attention on stevioside and if safety evaluations meet Western criteria and are successfully completed, then it could become a force to be reckoned with in the very competitive market for new sweeteners.

## 7. TALIN (THAUMATIN)

Talin sweetener is the trade name given by Tate & Lyle, PLC, to the mixture of intensely sweet-tasting proteins known as thaumatins which occur in the fruit of the West African plant *Thaumatococcus daniellii*. The botany, horticulture and physical properties of the thaumatins were reviewed comprehensively in the previous volume of this series,[78] but since 1979 considerable progress has been achieved in the production of Talin and its applications in a very wide range of food, drink and pharmaceutical products. Its known taste limitations as a complete sweetener have

* My Home Sugar Strong® is a registered trade mark of The Home Co., Japan.

encouraged developments in its special properties of flavour and aroma enhancement. The major obstacle for any new food additive, even a wholly natural one, is the necessity for toxicological evaluation and legislation/ registration. This has been largely overcome by the completion of safety data during 1981, allowing the gradual introduction of Talin as a commercially available product.

## 7.1. Production of Talin

The major source of raw material, the arils of *T. daniellii*, has shifted from Ghana and Nigeria to the French Ententes of Ivory Coast and Togo in West Africa. The natural abundance of *T. daniellii* fruit there is sufficient to supply needs for many years to come but in order to reduce the dependency for continued supply from any one country, particularly in politically sensitive areas of West Africa, other sub-tropical areas have been investigated. The Malaysian Peninsula has proved suitable for cultivation of *T. daniellii* using plants from both West Africa and the Singapore Botanic Gardens. Small *T. daniellii* plantations established as an intercrop with rubber have begun to produce fruit containing thaumatin. The plants are not yet mature enough to match West African levels but should provide a good alternative source of supply. The original plantation established in Bunso, Ghana, in 1976 continues to thrive and provides valuable data on plant and fruit development.

The arils are separated mechanically from the split fruit and extracted and purified to produce Talin. Only aqueous extraction and physical methods of separation and purification are employed to preserve the natural extract status of Talin.[79] Enough production capacity exists to supply the Japanese market and the developing applications in Europe and the USA.

The protein structure of thaumatin has proved an irresistible challenge to the genetic engineer, and several groups, notably Unilever in Vlaardingen, and Tate & Lyle in collaboration with the University of Kent, have made some progress. The Unilever group has reported[80] cloning double-stranded complementary DNA (converted from messenger RNA isolated from *T. daniellii* arils where thaumatin synthesis occurs naturally) into *Escherichia coli* HB101. The cloned DNA was 'engineered' to code for various proteins including a 'preprothaumatin'. This contained extensions to the normal thaumatin molecule, 22 amino acids with hydrophobic character at the amino terminus and 6 amino acids with acidic character at the carboxyl terminus. These are postulated to be 'signal' sequences which aid the transmission of proteins through membranes.

There are still many hurdles to be overcome before genetically engineered thaumatin is a practical proposition. It would be difficult to compete with aril tissue if 30 litres of bacterial culture were required to produce as much as one aril, or 100 litres for one whole fruit.[81] Microbial thaumatin has to be recovered, probably after modification by enzymes, then separated from cell components and other proteins and genetic material. The 'export' of proteins from cells has been closely studied, and basic research reported recently[82] suggests that a net positive charge on the signal protein helps it to pass through the cell membrane (thaumatin has 11 positive lysine groups) and that three regions in the protein are involved in protein 'export' including a 25-amino terminal signal sequence (prepro-thaumatin has 22-amino terminal amino acids).

A final problem is the attitude of regulatory authorities to foods, or food additives, produced genetically. The FDA is still undecided but will consider them in general as new products requiring some safety testing on the process itself and on the strains of microorganism used. The FDA is still awaiting the first formal submission to set a precedent for their policy for regulating such products.

Some attempts at tissue culture of parts of the *T. daniellii* plants were attempted in 1976. With the increased knowledge and techniques now available, several university groups have become interested in trying to make a relatively large molecule such as thaumatin.

### 7.2. Structure of the Thaumatins

The amino-acid sequences of the two main protein constituents of Talin, known as Thaumatin I and II, have been determined, and that of the main protein, Thaumatin I, has been published.[83] This shows a single chain of 207 amino acids with eight disulphide bridges giving a molecular weight of 22 209. This shows that the previously published amino-acid content derived from acid hydrolysis as usual underestimated the totals and is more correctly shown in Table 1.

Thaumatin II was studied by the Unilever Group and shows only five amino-acid differences, with a few ionic charge changes with rearrangements of the basic amino acids.[84]

The sequences studies show that the relative abundance of the basic amino acids, arginine (12 residues) and lysine (11 residues), plus the weakly basic asparagine (10 residues) and glutamine (4 residues), compared with the acidic amino acids aspartic (12 residues) and glutamic (6 residues) is responsible for the high overall isoelectric point of around pH 11.

TABLE 1

AMINO-ACID COMPOSITION OF THAUMATIN I $(T_1)^a$

| Amino acid | Number of residues | Percentage of total protein |
|---|---|---|
| Glycine | 24 | 6·2 |
| Threonine | 20 | 9·1 |
| Alanine | 16 | 5·1 |
| Half-cystine | 16 | 7·4 |
| Serine | 14 | 5·4 |
| Arginine | 12 | 8·5 |
| Aspartic acid | 12 | 6·2 |
| Proline | 12 | 5·2 |
| Lysine | 11 | 6·4 |
| Phenylalanine | 11 | 7·3 |
| Asparagine | 10 | 5·2 |
| Valine | 10 | 4·5 |
| Leucine | 9 | 4·6 |
| Isoleucine | 8 | 4·1 |
| Tyrosine | 8 | 5·9 |
| Glutamic acid | 6 | 3·5 |
| Glutamine | 4 | 2·3 |
| Tryptophan | 3 | 2·5 |
| Methionine | 1 | 0·6 |
| Total | 207 | 100 |

$^a$ Adapted from Iyengar et al.[83]

The lysine side-chain is unmodified in the native protein and, indeed, any modification which reduces its charge causes a reduction in sweetness.

A more striking feature of the thaumatins is the eight disulphide bridges which give a cross-linked network of amino-acid chains, conferring useful stabilities to heat and pH denaturation. Disulphide bridges are common in mobile proteins that operate in extracellular space such as digestive enzymes, immunoglobulins and milk proteins. The presence of Cys–Cys at positions 158 and 159 in the sequence is another frequently encountered structural element which forms the basis of linking three chain segments close together, again conferring stability. The disulphide bridges are also responsible for holding the protein chain in the correct conformation to elicit sweetness, as chemical cleavage of even one disulphide bridge results in a loss of taste.[85]

### 7.3. The Effect of Structural Modification on the Taste of Thaumatin

Reduction in the charge on lysine groups by acetylation progressively diminished sweetness whereas methylation did not. Additionally, modification of arginine's cationic guanidino group with cyclohexan-1:2-dione caused a net charge decrease but no taste change whereas drastic modification of carboxyl groups by esterification, or succinylation of amino groups, completely eliminated sweet taste.[78] A more recent report[86] showed that modification of the single methionine residue with iodoacetic acid caused no sweetness loss (but it did taste more like sucrose), whereas modification of tryptophan (three residues in Thaumatin I) by Koshland's reagent resulted in complete loss of taste. A more surprising change is an apparent *increase* in sweetness perceived after blocking the negative charge on carboxylic acid groups by carbodiimide-promoted amidation. Both the slow taste perception and the sweet aftertaste were much prolonged, combining to give an apparent sweetness some 12 000 times that of sucrose or some six times greater than normal thaumatin. However, linking of *secondary* amine groups onto the carboxyl groups causes progressive loss of taste. Finally, two of the seven tyrosine residues have been iodinated without loss of taste, but incorporation of three or more iodine atoms causes complete loss and drastic changes in the molecular backbone structure.

Another unexpected observation is that cleavage of one of the eight disulphide bonds not only causes loss of sweetness but converts it to a molecule possessing enzymic activity. Reduction with dithiothreitol causes rapid autodigestion to non-sweet peptides and amino acids and also results in weak protease, amidase and esterase activities. Acetylation of the lysine ε-amino groups leads to increased enzymic activity correlating with the number of acetyl groups introduced.[85]

All these observations confirm that both tertiary structure and ionic charge are important to sweet taste, with lysine probably involved in the binding of thaumatin to the taste receptors.

### 7.4. The Taste Quality of Thaumatin

The somewhat unusual sweet-taste profile of thaumatin (delay in sweet-taste perception followed by a lingering sweet taste) has been fully described previously.[87] Also the effects of pH, ionic strength, heat and the presence of other ingredients on taste perception were discussed, showing that, as with other sweeteners, these factors must be taken into account in food and drink applications.

Several Japanese formulations have set out to improve the taste profile of

Talin to make it more 'sucrose-like'. Combinations of thaumatin with glycyrrhizin, amino acids, citric and succinic acids, monosodium glutamate (MSG), and lactose were marketed by the San-Ei company as San-Sweet T-1®,* containing *circa* 2% thaumatin, with an overall sweetness 80–100 times that of sucrose. Sweetness is thereby enhanced. An improved formulation omitting glycyrrhizin is now marketed as San-Sweet T-100®,* with lower levels of amino acids (27% D-L-alanine) and MSG, but with a combination of tartrates, malates and succinates. Even without MSG a wide range of flavours is enhanced by this formulation.

Although aqueous solutions of these formulations taste more like sugar, the replacement of sucrose in soft drinks, for example, is still limited to around 50% before an obvious difference in sweetness quality is noted. However, this allows considerable calorie reduction while making use of the other major attribute of thaumatin, flavour enhancement.

Rats appear not to detect high concentrations of Talin, as their taste nerves show no electrophysiological response. However, in experiments in which rats are conditioned to reject *circa* 6% sucrose solutions but not 1% sugar or 0·02% Talin solution (equivalent to *circa* 20–30% sucrose), the mixture of the two acceptable solutions is now rejected like the 6% sucrose solution. This suggests that the mixture now tastes like sucrose to the rat with the Talin apparently strongly enhancing the weak, threshold sucrose taste.[88]

## 7.5. The Applications of Talin as a Flavour Enhancer
### 7.5.1. Japanese Products
Since the first chance observations that Talin enhanced the flavour of peppermint, very many practical applications have been developed, especially in Japan, to exploit this unusual property. Flavour enhancement is noted when Talin is used at concentrations below its sweet-taste threshold—typically at $5 \times 10^{-5}$% Talin. In Japanese formulations flavour enhancement and sweetness are achieved using $1$–$4 \times 10^{-4}$% Talin concentrations in the bulked products. Too many uses are described in product literature (some 42 items) to list all of them but flavour enhancement, cost reduction of sugars normally used and improved aroma are described for beverages, soft drinks, fruit juices, milk drinks, ice candies, sherbets, ice-creams, jellies, custards, chewing gum, wafers, soft candies, cream, pickles, fish pastes, sauces, cheeses, and dried fish.

* San-Sweet T-1® and San-Sweet T-100® are registered trade marks of San-Ei Chemicals Co., Japan.

Even more applications are described when Talin is combined with savoury flavour enhancers such as MSG and the 5'-nucleotides with, in addition, glycine, aspartate, tartrate and succinate. Besides general flavour enhancement of coffee, chocolate, milk, eggs, seaweed, fish products, etc., it is also claimed to enhance the aroma of shiso (soy sauce), crab flavouring, fruit essences and menthol. Sharp, spicy flavours, such as horseradish, mustard, ginger, garlic, onion and curry are made more 'appetite stimulating'. To complete the plethora of attributes, positive masking, rather than enhancement, of the more unpleasant side or 'raw' tastes is suggested for mutton, chicken, pork, sea urchin, kamaboko (boiled fish paste) and salted guts of squid. These claims may sound very broad but they have been shown to be effective in practice as Japanese companies have introduced many new products incorporating Talin as a flavour enhancer.

The use of Talin in Japanese chilled desserts is discussed in Reference 89 which emphasises the improvement of flavour, aroma and sweetness balance in reduced-calorie or low-cariogenic gelatin desserts.

The safety of any new food additive, whether natural or synthetic, is of paramount importance. To emphasise this aspect, the safety evaluation data available at that time were published[90] and voluntary standards were drafted for comment in the Japanese *Food Additive Federation News*,[123] which has responsibility for the safety of natural food additives. The introduction of similar products in the Western world still awaits official clearance as either a food additive or a flavour.

### 7.5.2. Western Food and Drinks

*7.5.2.1. Chewing gum.* When talin is used at 50–150 ppm the flavour of chewing gum—especially in the peppermint and spearmint types, and the coffee and the fruit flavours, such as lemon, plum and strawberry—is enhanced and prolonged. Also the sweetness and flavour are balanced throughout the taste period. Higher levels of Talin contribute some sweetness in sugarless gum and boost the relatively low sweetnesses of sorbitol, mannitol and xylitol used as bulk sweeteners, as well as lowering overall cariogenicity.[91]

*7.5.2.2. Coffee.* The pronounced enhancement of coffee flavour and aroma by Talin is still under investigation by the major coffee manufacturers in Europe and the USA. In our own laboratories, the addition of very low levels of Talin to instant coffee granules (made by redrying strong coffee solutions) enhances coffee strength by 10–15 %.

*7.5.2.3. Soft drinks.* The taste of Talin restricts its use in carbonated beverages to around 10–15 ppm, when it contributes up to 30% of the sweetness depending on the acidity and type of drink. However, certain flavours, such as blackcurrant, allow a higher replacement level (up to 50%) as the enhancement of flavour masks the aftertaste of Talin. A reduction of the sugar level in such a syrup drink (normally over 14% sucrose) not only reduces its calorie content and cariogenicity but also reduces 'stickiness' and improves flavour perception. Such levels also allow a reduction of 10–15% in the citric acid concentrations used.

Although the thaumatin molecule of Talin is stable to the low pH of soft drinks, its use has been limited by its ionic interactions with some soft-drink ingredients, such as synthetic colours (e.g. tartrazine) and caramel. This occasionally leads to cloudiness or eventual precipitation with gradual loss of sweetness. However, recently we have shown that the addition of weakly acidic gums, such as gum arabic, gives an overall neutral or slightly acidic charge, preventing these interactions and loss of sweetness and colour.

Finally, when Talin is used at much lower levels, i.e. 0·5–2·5 ppm, flavour is improved while the bitterness sometimes associated with saccharin-only-type drinks is masked.

*7.5.2.4. Flavour essences and oils.* The solubility of Talin in aqueous alcohol allows its direct addition to the flavour essence or oil providing, for example, a convenient route for its addition to chewing gum. Even more interesting, perhaps, is the recent observation (unpublished results, San-Ei Chemical Co. Osaka, Japan (1982)) that direct addition of Talin in 50% aqueous alcohol to the flavour essence results in enhanced *aroma* quality and persistence (Table 2). This is demonstrated in products where aroma is important—as in peppermint products such as mouthwashes and mint candies.

## 7.6. Applications in Medicines and Oral-Care Products
The intense sweet taste, non-cariogenicity and flavour/aroma-enhancement properties of Talin suggest useful applications in improving the palatability of medicines, particularly those with unpleasant-tasting long-lasting ingredients not easily masked by sugar or glucose syrups alone.

Applications which have been investigated include vitamin products, analgesics such as paracetamol, codeine derivatives, common cold remedies, antacids, cough syrups and oral antibiotic suspensions. The flavours used in these products, including peppermint, aniseed, cinnamon,

TABLE 2
THE EFFECT OF TALIN ON FLAVOUR AROMA

| Panel | Sample* | Wintergreen flavour | | Spearmint-oil flavour | | Lemon flavour | |
|---|---|---|---|---|---|---|---|
| | | 1 min† | 4 h† | 1 min† | 4 h† | 1 min† | 4 h† |
| 1 | a | Weak | Weak | Top strong | Powdery | Light | Fish |
| | b | Middle | Flavour of methyl salicylate | Middle | Middle | Middle | Middle |
| | c | Strong, heavy, sweet | Vanilla remains | Heavy, slow release | Strong, wide flavour | Citrus | Strong |
| 2 | a | Weak | Medium | Light, powdery | Weak | Weak | Weak, bad |
| | b | Middle | Equal | Middle | Middle | Middle | Middle |
| | c | Strong | Strongest | Strong, slow | Strong | Strong | Strong |
| 3 | a | Weak, medical smell | No good | Poor release | Weak | Top weak | Bad |
| | b | Middle | Strong | Good release | Middle | Harmony good | Quite strong |
| | c | Strong, sweet | Still strong | Heavy | Strong | Top strong | Strongest |
| 4 | a | Heavy | Methyl salicylate stays | Not sharp | Middle | Heavy, citrus | Weak |
| | b | Weak, not sweet | Weak | Rubbery, poor | Weak | Top strong, light | Middle |
| | c | Heavy, sweet | Strong, vanilla | Strong, good release | Strong | Heavy citrus | Strongest |

* a—Control; b—contains 50 ppm Talin added as a 25 mg ml$^{-1}$ Talin solution in 50% ethanol; c—contains 125 ppm Talin.
† Test time.

wintergreen, liquorice, orange, cherry, vanilla, banana, blackcurrant, lemon and herbal spices, are all accentuated by Talin. Levels of use can start as low as 0·005 % in liquid preparations where flavour enhancement and taste masking are the main requirements and can rise to 0·05 % or even higher when particularly strong tastes are present and strong sweetness is required.

Talin is also useful in some toothpaste and mouthwash formulations because of its non-cariogenicity and flavour enhancement. Peppermint, spearmint, cinnamon, wintergreen and cherry are particularly compatible with it, leaving a fresh, clean taste in the mouth. In general, the levels of use required can be as high as 0·1 % because of the relatively high concentrations of other ingredients which can mask the taste of Talin and, indeed, of other sweeteners/enhancers normally used.

### 7.7. Safety Evaluation and Legislative Status

It would not be expected that a natural plant protein of known structure containing normal amino acids would exhibit any toxicological problems. However, to demonstrate its safety for human use, Talin has, since 1976, undergone an extensive series of safety tests to enable it to be evaluated by the appropriate regulatory authorities. These tests have shown no adverse effects. The toxicological studies are described elsewhere (Chapter 6) and indicate a safety factor of at least 80 000 times the expected daily human intake level. As a result of these studies Talin has been permitted as a safe, natural food in Japan since May 1979. More recently (October 1981) the UK Committee on the Safety of Medicines has cleared Talin as a safe excipient in medicines, and in March 1982 the Committee on Toxicology recommended Talin as suitable for use in foods. This was endorsed by the Food Additives and Contaminants Committee in May 1982. Approval as a sweetener, along with aspartame and Acesulfame-K, is expected by mid-1983. In the USA, the position will soon be resolved regarding its status as a flavour enhancer. Petitions for clearance in Canada and Australia are well advanced and others will follow after favourable UK review. JECFA has placed Talin on their priority list and will review it in 1983.

## 8.  GLYCYRRHIZIN

Glycyrrhizin is better known as a liquorice-flavour additive and a medicinal taste masker than as a sweetener although it is approximately 50 times sweeter than sugar. Even the pure monoammonium glycyrrhizinate derivative (MAG) has a liquorice taste. The better-known and commercially widely available fully-ammoniated glycyrrhizin (AG) is on the US GRAS list as a natural flavour and is used as such in medicines, tobacco (to reduce harshness) and some confectionery products. However, it has the property of synergy with sucrose to give an apparent sweetness 100 times that of sucrose in a 50:50 mixture. The FDA has been petitioned to permit its use as a sweetness enhancer at a level of 0·4 % in chewing gum, 0·24 % in soft candies and 0·17 % in other foods. The FDA has not yet ruled on this but has indicated that if glycyrrhizin is affirmed as GRAS then it would only be for flavouring, not sweetening. Glycyrrhizin also has a well-known pharmacological activity and at high levels can cause oedema and hypertension. Concern about the high level of use in Japan led the Ministry of Health to issue cautionary statements in 1978 limiting its use in drugs to less than 200 mg per day.

There are practical limitations, too, as glycyrrhizin can precipitate from solution below pH 4·5, making it unsuitable for carbonated soft drinks except for the less acid, highly flavoured drinks such as root beer and cream soda.

### 8.1. Modifications to Taste and Structure of Glycyrrhizin
Several formulations claim to improve the properties of glycyrrhizin while minimising the liquorice aftertaste by using lactose and/or sorbitol, cream of tartar and a nucleotide.[92] A potassium–magnesium–calcium glycyrrhizin salt is claimed to have better solubility and no ammonia taste.[93]

The α-isomer of glycyrrhizin[94] has similar properties to the normal β-isomer but is more soluble and stable. Apart from its sweetness, it is claimed to be a surfactant, an anti-viral, anti-ulcer, anti-tumour, anti-spasmodic and tissue-repair agent, and a sex hormone! These extra properties do not bode well for its use as a sweetener! Extraction of liquorice with strong hot alkali is claimed to give a glycyrrhizin that does not gel in acid and which has a quoted sweetness 100–300 times that of sugar.[95]

Thus it is most unlikely that glycyrrhizin will be widely used as a sweetener, even when formulated with flavour modifiers. It will continue in use in many products, especially chewing gum and chocolate confectionery, in which, although restricted in its use as a flavour, it does contribute significantly to the product's sweetness.

## 9.   THE DIHYDROCHALCONES

The group of flavonoid derivatives known collectively as the dihydrochalcones have been investigated for nearly 20 years, but are still not used widely. The best known, neohesperidin dihydrochalcone (NeoDHC), is generally quoted as being 2000 times sweeter than sugar but is, in fact, some 350 times sweeter than sucrose at $8.5\%$ concentration. Despite this still encouraging sweetness level and adequate stability in actual formulations, it does have the major shortcomings of a delay in perception of the sweet taste and a long sweet and slightly bitter liquorice aftertaste, with a pronounced menthol-like cooling effect on breathing-in while tasting. Other limitations are its low water solubility ($1.2$ g litre$^{-1}$), instability of the carbohydrate residues to acid conditions leading to slow loss of sweetness,[96] and occasional formation of a yellow discoloration on standing in aqueous solution,[97] which, it is claimed, can be avoided.

## 9.1. Taste Improvements to the Dihydrochalcones

### 9.1.1. Synthetic Derivatives

The less than optimum sweetness quality of NeoDHC prompted several groups of workers to search for better-tasting synthetic alternatives, of which at least 250 variants were prepared, none of which was ideal, but a lot was learnt about the structure–sweetness relationship. A comprehensive search for a hypothesis relating structure to perception delay and lingering aftertaste showed[98] that none of the four variables—metabolic, conformation, chelation or hydrophobicity—was responsible and none of the 44 analogues had a better taste. It was known that the carbohydrate groups are not necessary for sweetness and could be replaced by hydrophilic side-chains of greater stability to hydrolysis such as sulphoalkyl and carboxyalkyl with and without amino groups.[68] Taste quality was little improved although less intestinal absorption resulted. Other larger molecular weight conjugates were prepared to reduce or prevent absorption and to demonstrate the Dynapol philosophy that sweet, polymeric, non-absorbable molecules were possible.[99] The first real improvement was claimed[100] for the homoserine dihydrochalcone conjugate, which had similar sweetness ($400 \pm 30$ times that of sucrose) but significantly less persistence. It was suggested that less binding to the back part of the tongue was occurring, as it was this that was thought to be responsible for the lingering aftertaste. Certainly, long-lasting sweet-tasting substances such as glycyrrhizin, thaumatin and monellin, as well as NeoDHC, all initiate taste responses at the front and then linger at the back of the palate.

### 9.1.2. NeoDHC Formulations

Another approach to taste improvement is to formulate mixtures with taste modifiers and bulking agents. The addition of cream of tartar, a carbohydrate bulking agents and vanilla flavour was claimed to eliminate perception delay and aftertaste.[101] The addition of gluconates,[102] amino acids[103] or nucleotides[104] has also been claimed to improve sweetness quality.

### 9.2. Some Uses of NeoDHC in Foods and Drinks

The long-lasting sweetness of NeoDHC has generally recommended its use where such a property is desirable, such as in chewing gum, bitter-tasting fruit juices, toothpastes, mouthwashes and some medicines. Use in grapefruit juice,[105] to mask bitter naringin tastes, is particularly recommended as is use in some bitter-flavoured orange juices such as that of the Californian navel orange.

Up to 25% of the sweetness in soft drinks can be contributed by NeoDHC before the taste becomes unacceptable.[96] Occasional interaction with other ingredients has been found to lead to precipitation in orange-juice formulation.[106]

The flavour of NeoDHC is said to be incompatible with tea and coffee[107] although Dynapol claims[108] that the aminocarboxyl derivative is suitable for such beverages. The non-cariogenicity of NeoDHC is the basis for its suggested use in gum and dentrifrices in conjunction with sugar alcohols and glycerol.[109] Also in the USA, Nutrilite Products Inc., one of the main producers of NeoDHC, claims[110,111] that the aroma and flavour in chewing gum is preserved by low levels of addition. More unusual applications are for sweetening cultured milks, especially raspberry yoghurt,[112] suppressing salt taste in highly brined traditional Japanese foods,[113] and as a tobacco flavourant.[114]

### 9.3. Future Marketing Prospects for NeoDHC

Israeli, American and Spanish companies with access to large quantities of citrus raw materials continue to develop the manufacture of NeoDHC and are pressing for regulatory clearance. Jaf-Ora offer development quantities at around £700 per kilogram, or five times the price of the sucrose equivalent. In 1975,[115] Nutrilite hoped to sell NeoDHC at US$75–125 per pound, close to the price of sucrose in use, and had a reported 25-ton capacity.

The practical problems of using NeoDHC are compounded by the difficult regulatory position. Despite many years of safety testing by American companies and the USDA, the FDA refused to give NeoDHC GRAS status and deemed that the existing toxicological data were inadequate to support a food-additive petition. Most of the studies were performed under guidelines which do not conform to present FDA Good Laboratory Practice and will, therefore, need to be repeated. The cost of this and the fact that the USDA, as a Government agency holding patents on NeoDHC, cannot issue exclusive licences means there is little incentive for individual companies to market this product. US companies such as Atomergic offer NeoDHC as a research chemical, and interests in Spain suggests that there may be a limited market in some European countries where it may be approved on the evidence of the existing safety studies. A recent note on pending Belgian food-additive legislation proposed that NeoDHC be allowed in certain beers, chewing gum and lemonades at 20 mg litre$^{-1}$ as a saccharin replacement, but this has not yet been implemented. Clearly, NeoDHC still has many hurdles before it can achieve a limited market as a specialised sweetener.

## 10.  SOME NEWER SYNTHETIC SWEETENERS

Many laboratories throughout the world have an active programme for synthesising novel sweet compounds, and all are looking for a safe, high-intensity, easily produced, inexpensive sweetener of good taste quality which is stable in use. Four recently patented sweeteners meet some of these criteria. Two are derivatives of aspartic acid: Procter & Gamble's L-1-hydroxymethylalkylamide[116] and Pfizer's L-aspartyl-D-alanine-N-(dicyclo propylcarbinyl)-amide.[117] Eli Lilly describes a 5-(2,3-dihydroxyphenoxy)-tetrazole, with a sweetness threshold 1200 times that of sucrose (considerably less at actual use levels), with some aftertaste but sweetness synergy with saccharin and cyclamate.[118] Little is known about its safety evaluation. Pfizer's product is similarly sweet, apparently with good taste quality, synergy with saccharin and greater stability under acid conditions than aspartame. The Procter & Gamble derivatives of aspartic acid are less potent at 50 times the sweetness of sucrose, but are more stable than aspartame as they lack the unstable ester linkage.

The fourth group of new sweeteners is that of the chloro-derivatives of sucrose, of which the trichloro-derivative has been selected for further development.

### 10.1.  Trichlorogalactosucrose (TGS)

Sucrose as a relatively cheap, abundantly available, pure raw material with its many hydroxyl groups, is an ideal target for chemical structural modification to find sweeter derivatives. Only slightly sweet compounds, such as the monoesters or the C-4 epimer *galacto*sucrose, or very bitter derivatives, such as the octa-acetate, were known before 1976, when Hough and Phadnis,[119] in collaboration with Tate & Lyle, discovered the intense sweetness of 4,6,1′,6′-tetra-chloro-*galacto*sucrose. Further chlorodeoxy-derivatives were synthesised to determine the effect on taste of substituting the different hydroxyl groups of sucrose by chlorine (1943 derivatives are theoretically possible). For example, it was found that the 6-chlorodeoxy-derivative is bitter, the 4-chlorodeoxy*galacto*-derivative some five times as sweet as sucrose and the 4,1′,6′-trichloro-derivative extremely sweet. The latter compound, 4,1′,6′-trichloro-4,1′,6′-trideoxy*galacto*sucrose (known as TGS), was found to be about twice as sweet as saccharin, but had a taste almost indistinguishable from that of sucrose at normal concentrations with no bitter or metallic aftertaste. This discovery was so surprising and important that world-wide patent coverage was immediately sought.

TGS is obtained as a pure, white crystalline compound, molecular weight 397·64 ($C_{12}H_{19}O_8Cl_3$), readily soluble in water and the lower alcohols and other polar solvents to give a clear colourless solution of neutral pH. The accurate measurement of sweetness is subjective and affected by many variables such as concentration, temperature, pH, ionic strength and the presence of other ingredients. At room temperature in water TGS has a relatively flat sweetness/concentration curve compared with other sweeteners (Fig. 1) giving a value of around 500–600 relative to sucrose at a concentration of 6–8 %. Just as important as the quality of sweet taste is the taste–time profile. TGS scores highly, as there is an almost imperceptible delay in the onset of sweetness and only a slight lingering sweet aftertaste, less noticeable than with aspartame. In practical applications such as soft drinks neither of these differences from sucrose is perceptible. Thus of all the known sweeteners, TGS is generally considered to have a sweet taste closest to the ideal—that of sucrose itself.

It is not enough, however, to have a good taste in water. The sweetener must be stable enough to withstand the conditions of food processing and to remain unchanged over a wide range of pH, especially the low pH conditions of some soft drinks, such as colas, and over a long shelf-life at varying temperatures. Furthermore, it should not react with other ingredients of food or drinks. TGS is found to be particularly stable in solution at low pH and elevated temperatures, being appreciably more resistant to hydrolysis than its parent sucrose. The chlorine substituents also render TGS resistant to enzymic hydrolysis, especially to invertase and $\beta$-fructofuranosidases, so that TGS is not metabolised by digestive enzymes. Neither is it a substrate for oral bacteria and thus does not contribute to

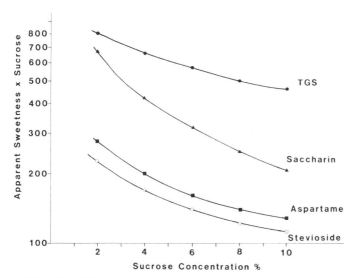

FIG. 1.   The effect of increasing sucrose equivalent concentration on the apparent sweetness factor of high-intensity sweeteners.

dental caries. TGS should be suitable for diabetic use, as at sucrose replacement levels it has no effect on insulin secretion.[120] There are no observed interactions with typical food ingredients such as proteins, gums, other carbohydrates or tannins. The heat stability of TGS is similar to that of sucrose, inferring that it can be used in normal cooking and baking. TGS thus has the attributes of an ideal high-intensity sweetener. To prove its safety in use, a comprehensive series of toxicological tests *in vivo* and *in vitro* is being undertaken as a pre-requisite for seeking regulatory approval as a safe food additive. Only then will TGS be able to find its place in dietary products and extend the choice of the high-intensity sweeteners now available.

## 11.   CONCLUSIONS

Despite all the effort, expenditure and determination of many years of research and development, no single ideal substitute for sucrose has yet been found which can completely match its qualities of sweet taste, wide range of applications and sweetness stability. The calories contributed by sucrose can be reduced but only by compromising some sweetness quality and occasionally stability and safety, and then only when water is the major bulking agent as in soft drinks. As more alternatives become available, blends of nutritive and non-nutritive sweeteners will be useful, and mixed high-intensity sweetener combinations, if permitted on safety grounds, could be ideal in specialised products, especially diet beverages, pharmaceuticals, oral-care products and where a novel or extended flavour is required such as in chewing gum.

For safety reasons, a range of available sweeteners is welcomed as it reduces the intake of any one particular sweetener and spreads any possible risk involved. The food industry also benefits by having greater freedom to select a particular sweetener for a specialised use to bring out its individual advantages. The challenge now is to the food formulator to select innovative combinations to produce better-tasting, higher quality and more appealing foods for tomorrow's consumer.

## REFERENCES

1. FURIA, T. E. (1968). *Handbook of Food Additives*, 1st edn. CRC Press, Cleveland.
2. *Food Additives and Contaminants Committee Report on the Review of Sweeteners in Food* (1982). 10th March, HMSO, London.
3. G. D. SEARLE & CO. (1967). British Patent 1 152 977.
4. IMPERIAL CHEMICAL INDUSTRIES LTD. (1965). British Patent 1 042 484.
5. AJINOMOTO CO. INC. (1970). British Patent 1 243 169.
6. AJINOMOTO CO. INC. (1971). British Patent 1 309 605.
7. MONSANTO CO. (1973). German Patent 2 452 285.
8. MONSANTO CO. (1976). German Patent 2 757 771.
9. STAMICARBON N. V. (1972). British Patent 1 359 123.
10. BEECHAM GROUP LTD. (1971). German Patent 2 107 411.
11. ANON. (1981). *Biotechnology News*, 1(17), 1st September.
12. TOYO SODA MANUFACTURING CO. LTD. (1978). US Patent 4 165 311.
13. G. D. SEARLE & CO. (1981). US Patent 4 293 648.
14. CETUS CORPORATION (1981). European Patent 36258.
15. G. D. SEARLE & CO. (1981). Netherlands Patent Appl. 8 100 437A.
16. GENERAL FOODS CORPORATION (1974). British Patent 1 508 662.
17. GENERAL FOODS CORPORATION (1974). British Patent 1 462 799.
18. GENERAL FOODS CORPORATION (1974). British Patent 1 468 222.
19. GENERAL FOODS CORPORATION (1974). British Patent 1 468 071.
20. PROCTER & GAMBLE CO. (1975). British Patent 1 535 415.
21. ALBERTO CULVER CO. (1974). US Patent 4 982 98. (Re-issue of 3 753 739 (1970).)
22. GENERAL FOODS CORPORATION (1976). British Patent 1 569 636.
23. GENERAL FOODS CORPORATION (1974). Canadian Patent 1 046 840.

24. G. D. SEARLE & CO. (1970). Dutch Patent 16 19 78.
25. SEARLE BIOCHEMICS TECHNICAL BULLETIN, No. 600. *Aspartame*.
26. GENERAL FOODS CORPORATION (1973). Canadian Patent 1 028 197.
27. GENERAL FOODS CORPORATION (1973). Canadian Patent 1 026 987.
28. COLGATE PALMOLIVE CO. (1975). US Patent 4 071 615.
29. SUNSTAR HAMIGAKI CO. (1979). Japanese Patent 084 800.
30. GENERAL FOODS CORPORATION (1981). US Patent 4 277 464.
31. GENERAL FOODS CORPORATION (1979). Belgian Patent 882 672.
32. GENERAL FOODS CORPORATION (1977). US Patent 4 139 639.
33. GENERAL FOODS CORPORATION (1972). US Patent 4 036 992.
34. MCCORMICK, R. D. (1975). *Food Product Dev.*, **9**(1), 22.
35. GENERAL FOODS CORPORATION (1981). Japanese Patent 8 149 542.
36. BALDWIN, R. E. and KORSCHGER, B. M. (1979). *J. Food Sci.*, **44**, 938.
37. CUMBERLAND PACKING CORPORATION (1974). British Patent 1 465 728.
38. CUMBERLAND PACKING CORPORATION (1974). British Patent 1 466 096.
39. CUMBERLAND PACKING CORPORATION (1978). US Patent 4 254 154.
40. GENERAL FOODS CORPORATION (1976). US Patent 4 143 170.
41. AJINOMOTO, K. K. (1980). Japanese Patent 051 383.
42. NABISCO INC. (1976). British Patent 1 538 390.
43. GENERAL FOODS CORPORATION (1976). British Patent 1 511 392.
44. GENERAL FOODS CORPORATION (1972). US Patent 3 829 588.
45. AJINOMOTO, K. K. (1980). Japanese Patent 035 046.
46. HUGILL, J. A. C. (1982). *World Sugar Res. Org. Newsletter*, **16**, March, 2.
47. ANON. (1981). *Eur. Chem. News.*, July 27, 15.
48. QUOTED BY ANON. (1981). *Food Manufacture*, November, 19.
49. ANON. (1982). *Biotechnology News*, April 15, 2; **8**, 4.
50. CLAUSS, K. and JENSEN, H. (1973). *Angew. Chemie*, **85**, 965.
51. CLAUSS, K., LÜCK, E. and VON RYMAN LIPINSKI, G.-W. (1976). *Z. Lebensm. Unters-Forsch.*, **162**, 37.
52. HUDDART, W. (1982). Personal communication, Hoechst, UK.
53. ANON. (1982). *Food Chem. News*, March 22, 7.
54. VON RYMAN LIPINSKI, G.-W. and LÜCK, E. (1981). *Manuf. Chemist Aerosol News*, May, 37.
55. LÜCK, E. (1981). *Ann. Fals. Exp. Chim.*, **74**(796), 293.
56. TAKASAGO PERFUMERY CO. (1973). Japanese Patent 136 217.
57. MARUZEN CO. (1980). Japanese Patent 009 940.
58. TOYO SEITO CO. (1980). Japanese Patent 040 310.
59. AJINOMOTO CO. (1980). Japanese Patent 023 757.
60. AJINOMOTO CO. (1980). Japanese Patent 023 756.
61. OGAWA, T., NOZAKI, M. and MATSUI, M. (1980). *Tetrahedron*, **36**(18), 2641.
62. KASAI, R. *et al.* (1981). *Nippon Kagaku Kaishi*, **5**, 726.
63. DUBOIS, G. E., DIETRICH, P. S., LEE, J. F., MCGARRAUGH, G. V. and STEPHENSON, R. A. (1981). *J. Med. Chem.*, **24**(11), 1269.
64. WADA, Y., TAMURA, T., KODAMA, T., YAMAKI, T. and UCHIDA, Y. (1981). *Yukagaku*, **30**(4), 215.
65. OHASHI, S. (1981). Personal communication, San-Ei Co., Japan.
66. CHENG, T. F., CHANG, W. H. and CHANG, T. R. (1981). *K'o Hsueh Fa Chan Yueh K'an.* **9**(9), 775.

67. YOSHIKAWA, S., ISHIMA, T. and KATAYAMA, O. (1979). *Amer. Chem. Soc. Abstr.*, **177**(1) AGFD, 74.
68. CROSBY, G. A. and WINGARD, R. E. (1979). In: *Developments in Sweeteners—1*, C. A. M. Hough, K. J. Parker and A. J. Vlitos (Eds.), Applied Science Publishers Ltd, London, 135.
69. AJINOMOTO CO. (1979). Japanese Patent 8 111 772.
70. OGONTO CO. (1979). Japanese Patent 8 155 174.
71. UENO PHARMACEUTICAL CO. (1980). Japanese Patent 062 690.
72. NISSHIN SUGAR CO. (1981). Japanese Patent 81 131 400.
73. JAPAN TOBACCO & SALT PUBLIC CO. (1981). Japanese Patent 059 573.
74. JAPAN TOBACCO & SALT PUBLIC CO. (1981). Japanese Patent 059 574.
75. HAYASHIBARA CO. (1978). British Patent Application GB 2 027 423A.
76. MARUZEN CO. (1979). Japanese Patent 117 933.
77. ANON. (1981). *Food Chem. News*, May 11, 43.
78. HIGGINBOTHAM, J. D. (1979). In: *Developments in Sweeteners—1*, C. A. M. Hough, K. J. Parker and A. J. Vlitos (Eds.), Applied Science Publishers Ltd, London, 87.
79. TATE & LYLE PLC (1979). European Patent 30911.
80. EDENS, L. *et al.* (1982). *Gene*, **18**, 1.
81. ANON. (1980). *The Times*, December 30, Science Report.
82. ANON. (1982). *Genetic Technol. News*, May, 5.
83. IYENGAR, R. B. *et al.* (1979). *Eur. J. Biochem.*, **96**, 193.
84. VAN DER WEL, H. Personal communication, Unilever Research, Vlaardingen, The Netherlands.
85. VAN DER WEL, H. and BEL, W. J. (1980). *Eur. J. Biochem.*, **104**, 413.
86. VAN DER WEL, H. (1982). *Olfaction and Taste VII*, 13.
87. HIGGINBOTHAM, J. D., LINDLEY, M. G. and STEPHENS, J. P. (1981). In: *The Quality of Foods and Beverages*, G. Charalambous and G. Inglett (Eds.), Academic Press, New York.
88. DU VILLARD, X. D., VAN DER WEL, H. and BROUWER, J. N. (1980). *Chemical Senses*, **5**(2), 93.
89. OCHI, T. (1981). *New Food Ind.*, **23**(2), 13.
90. OHASHI, S. (1981). *Food Sanitation Research*, **31**, 9.
91. LEACH, S. A. and GREEN, R. M. (1981). *Caries Res.*, **15**, 508.
92. MACANDREWS & FORBES CO. (1978). German Patent 2 851 082.
93. MACANDREWS & FORBES CO. (1978). US Patent 4 176 228.
94. MARUZEN CO. (1980). Japanese Patent 024 023.
95. MARUZEN CO. (1979). Japanese Patent 172 494.
96. INGLETT, G. E., KRBECHEK, K. L., DOWLING, B. and WAGNER, R. (1969). *J. Food Sci.*, **34**, 101.
97. CHIMICASA GMBH. (1981). US Patent 4 254 155.
98. DUBOIS, G. E., CROSBY, G. A. and STEPHENSON, K. A. (1981). *J. Med. Chem.*, **24**, 408.
99. BROWN, J. P. *et al.* (1978). *J. Agric. Food Chem.*, **26**(6), 1418.
100. DUBOIS, G. E., CROSBY, G. A., LEE, J. F., STEPHENSON, R. A. and WANG, P. C. (1981). *J. Agric. Food Chem.*, **29**, 1269.
101. CUMBERLAND PACKING CORPORATION (1976). US Patent 4 085 232.
102. GIVAUDAN CO. (1974). Swiss Patent 592 418.

103. TAITO CO. (1980). Japanese Patent 8 029 664.
104. ASAHI CHEMICAL INDUSTRY CO. (1980). Japanese Patent 8 046 699.
105. PRATTER, P. J. (1980). *Perfumer and Flavorist*, **5**(7), 1.
106. SUNSTAR DENTIER CO. (1980). Japanese Patent 086 535.
107. O'BRIEN, L. and GELARDI, R. (1981). *Chemtech.*, May, 274.
108. DYNAPOL INC. (1979). US Patent 4 227 165.
109. PROCTER & GAMBLE CO. (1971). Japanese Patent 07 67 58.
110. NUTRILITE PRODUCTS INC. (1971). German Patent 2 155 321.
111. NUTRILITE PRODUCTS INC. & WARNER LAMBERT CO. (1974). US Patent 3 857 962.
112. HUBER, U., KOSSLAKOFF, N. and VATERLAUS, B. (1977). US Patent 4 001 453.
113. TOREY CO. (1973). Japanese Patent 6 38 19.
114. TORAY INDUSTRIES LTD (1980). Japanese Patent 8 045 184.
115. SELTZER, R. J. (1975). *Chem. and Eng. News.*, August 25, 27.
116. PROCTER & GAMBLE CO. (1978). European Patent 1 30 44.
117. PFIZER INC. (1980). European Patent 3 48 76.
118. ELI LILLY & CO. (1980). European Patent 3 49 25.
119. HOUGH, L. and PHADNIS, S. P. (1976). *Nature*, **263**, 800.
120. NICOL, W. M. Tate & Lyle Food & Sweeteners Development Group, Reading.
121. OSER, B. L. (1975). *Toxicology*, **4**, 315.
122. JOINT FAO/WHO EXPERT COMMITTEE ON FOOD ADDITIVES. (1981). *Twenty-fifth Report*, WHO, Geneva, Techn. Rep. Series, **669**.
123. ANON. (1980). *Food Additive Federation News*, **57**, 5702.

*Chapter 6*

# THE TOXICOLOGY AND SAFETY EVALUATION OF NON-NUTRITIVE SWEETENERS

D. J. SNODIN

*Tate & Lyle Group Research and Development,
Reading, Berkshire, UK*

and

J. W. DANIEL

*Life Science Research, Stock, Essex, UK*

SUMMARY

*The safety assessment of non-nutritive sweeteners and other food additives relies largely upon empirical studies in laboratory animals. Several non-sugar sweeteners have been evaluated extensively for their safety-in-use although aspects of the toxicological profile of some await clarification. Differing interpretations of the experimental data result in the hetero-geneous pattern of availability of non-nutritive sweeteners around the world.*

'Anybody who says saccharin is injurious is an idiot'
President T. Roosevelt, 1907[1]

## 1. INTRODUCTION

The use of sweeteners other than sugar in the human diet has increased dramatically in the last decade and while only saccharin and cyclamate

157

command significant use world-wide, the dipeptide $N$-$\alpha$-aspartyl-L-phenylalanine methyl ester (aspartame) has been approved by many national authorities for use in various commodities such as powdered soft-drink mixes, dessert mixes, chewing gum and water ices.[2] Thaumatin,[2] a mixture of intensely sweet proteins isolated from the arils of *Thaumatococcus daniellii* will soon be permitted in the UK, as will Acesulfame-K®,*[2,3] the potassium salt of 6-methyl-1,2,3-oxathiazin-4(3H)-one-2,3-dioxide.

A number of other sweet substances are at various stages of development including, for example, miraculin, monellin, stevioside, $\beta$-neohesperidin dihydrochalcone and trichloro*galacto*sucrose.

The attribution 'non-nutritive' in the title of this chapter is somewhat misleading for in the US it is applied to any sweet substance having less than 2 % of the caloric value of sucrose per equivalent unit of sweetening capacity. This implies that any compound with the same caloric value as sucrose will be classified as 'non-nutritive' providing it is at least 50 times sweeter than sucrose. This terminology has not found universal support, and is replaced in the UK by 'artificial', which refers to 'any chemical compound that is sweet to the taste but does not include any sugar or polyhydric alcohol'.[4] Sugar alcohols, sometimes called bulk sweeteners,[5] are generally not sweeter than sucrose but often contribute fewer calories because of either poor absorption or utilisation. For present purposes the term non-nutritive sweetener (NNS) will be used to describe substances considerably sweeter than sucrose, without regard to any legal criteria.

## 2. SAFETY ASSESSMENT TECHNIQUES

The following section can only be an introduction to the complexities of toxicity testing and safety evaluation. Many excellent accounts of general principles and techniques and of special considerations for food additives can be found elsewhere.[6-17] Definitions of some relevant terms are shown in Table 1.

Nearly a century ago Dr Harvey W. Wiley became Chief of Chemistry at the US Department of Agriculture. Wiley put in hand various investigations, especially into food preservatives including saccharin which found use as a preservative, culminating in the setting-up of the 'poison

---

* Acesulfame-K® is a registered trade mark of Hoechst AG, West Germany.

TABLE 1
DEFINITIONS OF TOXICITY, HAZARD, RISK AND SAFETY

*Toxicity*—the capacity of a substance to cause injury to a living organism. In practice, it has to be qualified with respect to the size of the dose, route and time distribution of the dose.[16]

*Hazard*—the likelihood that a chemical will cause an adverse health effect under the conditions of (human) exposure.[16]

*Risk*—a statistical concept assessing the expected frequency of adverse health effects arising from exposure to the substance in question.[16]

*Safety*—the converse of risk, which should really be judged in terms of a specified, scientifically and/or socially acceptable low risk. The NAS/NRC definition[13]—that 'safety' is the practical certainty that injury will not result from the substance when used in the manner and the quantity proposed for its use—is of little use unless practical certainty is defined in some way.

squad'.[18] Volunteers took all of their meals at the 'hygienic table' consuming increasing amounts of the food chemical in question. Food and water intake, body weight, pulse rate, temperature, urine contents, blood count, etc., were carefully monitored, and the amounts of test substance excreted in urine and faeces were recorded. On the basis of such tests Wiley recommended to President Theodore Roosevelt that saccharin should be banned because it was 'highly injurious to health'. However, the President ignored this advice, claiming that his doctor gave saccharin to him every day.[1]

## 2.1. Present Procedures

### 2.1.1. General Considerations

There have been many changes since Wiley's time with the emphasis on studies in animals, supplemented more recently by the use of systems *in vitro*. However, before describing the principles of toxicity testing, it is worthwhile discussing various factors that would have to be considered before embarking upon any substantial programme of such studies on a particular NNS.

Data on *chemical structure and physicochemical properties* may permit some degree of prediction of toxicological structure/activity relationships. Toxicological data-bases are now readily available and computerised structure/activity systems are being developed.[19,20]

Conducting toxicological studies on samples of NNS drawn from a large homogeneous batch made using the same chemical plant and process that would be used eventually to manufacture the material for sale would be ideal. Unfortunately, it is often difficult to achieve. An adequate amount of pilot synthesis will enable an impurity profile and batch-to-batch variation to be established, from which a *specification* for the NNS may be drawn-up which, hopefully, will be met on the full-size plant.

*Purity standards* for existing NNS's are becoming more exacting due in part to the capability of modern analytical techniques to detect smaller and smaller amounts of impurities. Therefore, for a new NNS a pragmatic approach must be adopted in order to avoid endless analysis and toxicological testing of impurities. A combination of three concepts can be helpful:

1.  A limit figure for single impurities known to be present;
2.  Searching for impurities and by-products which may be formed and which could on theoretical grounds have toxic potential; and
3.  Accepting that all other impurities, although not identified, will form part of the material which is toxicologically evaluated. There may well be a case for separate toxicological evaluation of an unavoidable impurity.

*Specifications* for NNS's and other food additives, such as those listed by the Food Chemicals Codex and European Economic Community, concentrate on key parameters affecting identity, purity and safety. They are compiled when all or most of the NNS development (including toxicology) has been completed and so will benefit from hindsight as to the significance of the nature and concentration of particular impurities.

To paraphrase one of the earliest toxicologists (Paracelsus)—the dose defines the poison.[21] Thus the hazard to human populations consuming a particular NNS equals toxicity multiplied by intake. It is important to obtain an estimate of the likely *consumer intake* of an NNS at an early stage of development in order to ensure that there will be an acceptable margin of safety.

The three elements required for reliable intake estimates are:

1.  Concentration in each dietary item;
2.  Amount of each item consumed; and
3.  Frequency of eating relevant items.

Reliable data on point 1 are relatively easy to provide whereas on point 2 it is extremely difficult first to decide how large a particular food category should be, and second to cater for the great variability of individual consumption patterns. The second consideration also applies to point 3. Since, to err on the side of caution, it has become customary to assume that a particular NNS will be used in all food items comprising a particular category, it is accepted that intakes are often overestimated. Moreover, using the food-intake level of the 90th percentile has become customary as a conservative but reasonable figure for the entire population, again overestimating consumption by the average consumer.

A reasonable starting point for intake calculations involves the use of data presented in the review of the GRAS list by the US National Academy of Sciences–National Research Council (NAS–NRC) Committee on Food Protection,[22] where, *inter alia*, intakes of each of 28 food categories are listed for both 'eaters' and 'non-eaters' at the normal and 90th percentile consumption level. More specific data relating to commodities like soft drinks[23] will often be useful.

Several authors[14,24] have made critical assessments of commonly used methods for estimating consumer intake; each method has its particular strengths and weaknesses but may be hampered by the paucity of appropriate data.

Various *toxicological testing procedures* have been developed. The names of the studies reflect the size of the dose employed and the length of exposure to the test substance, moving from high dose/short-term (acute) to low dose/long-term (chronic) studies. The primary objectives of toxicity tests are:

1.    To determine the effects of the test material on biological systems and;
2.    To deduce the dose–response characteristics for these effects.

The species most commonly used are the rat, mouse and dog, with the rabbit, guinea-pig, hamster and non-human primate being used for specific purposes. Normal protocols utilise a control group of animals fed the test substance vehicle and at least three test groups fed graded doses of the test substance. During and after treatment, animals are observed for any signs of toxic response. Test- and control-group animals are subjected to laboratory procedures including haematology, blood biochemistry and urinalysis during the study. Bodyweights and food and water intake are also measured. After the specified treatment period the animals are killed and subjected to a pathological examination (both macroscopic and microscopic). Data are analysed primarily for dose-related group trends with respect to both biological and statistical significance.

## 2.2. Acute and Sub-Acute Toxicity Tests

Acute toxicity has been defined as 'the adverse effects occurring within a short time of administration of single or multiple doses given within 24 h'.[16] Nowadays the term is used more generally to describe tests employing high doses and/or conducted over short periods.

The most frequently quoted test is that used to determine the median lethal dose ($LD_{50}$) of the test substance, which was defined by Gehring as 'a statistically derived expression of a single dose of material that can be expected to kill 50 % of the animals'.[16] The test consists of treating groups of animals with a mathematically related series of doses, observing signs in treated animals for up to 7 or 14 days, and noting deaths. The determination of precise $LD_{50}$ values requires the use of large numbers of animals and amounts of compound, and the necessity for conducting such tests has been questioned.[25] The use of small numbers of animals in a 'limit test' will provide a good indication of the toxicity of the substance.

The objective of sub-acute tests is to provide frequent, daily or continuous exposure of the test substance to groups of animals over periods of up to 90 days in order to identify the major toxic effects, the target organs, and sometimes to investigate the reversibility of such effects.

The classic test is the 90-day rat test, details of which may be found in various texts.[6–10,16] Doses are set with reference to acute studies, palatability and metabolic characteristics, but not made so high as to produce excessive mortality or gross dietary imbalance. Normally, the highest dose is chosen in the expectation of producing some toxic effects. On completion, data from the test will be used to establish a 'no-effect level' (NEL) which is the dose that produces no toxicologically significant effect.

The traditional procedures have been criticised in that little insight is gained into the mechanisms of toxicity, and the origins, progression and reversibility of lesions. However, a well-conducted test incorporating, for example, reversibility and metabolic investigations, can be of considerable value. The use of tests of shorter duration, for example 28 days, is gaining acceptance since it has been argued that some of the objectives of a 90-day study can be satisfied in the shorter time span.[26] This view has not found universal acceptance in food-additive toxicology.

## 2.3. Chronic Toxicity and Carcinogenicity Studies

Chronic toxicity tests are conducted over the life-span of test animals at appropriate dosages in order that an assessment can be made of the likely toxicity in man on long-term low-level exposure. Since tumour development is no more than one manifestation of chronic toxicity, it is customary to submit food additives to a combined chronic toxicity and carcinogenicity study if this is indicated following an assessment of the exposure and sub-chronic data.

Species most commonly employed are the rat (24 or 30 months) and the mouse (18 or 24 months). Choice of dose levels is often difficult; making the top dose the maximum tolerated dose[27] (MTD—the dose that in sub-chronic studies did not produce a reduction in weight of more than 10% compared with controls), or some high proportion of it, may cause nutritional imbalance such as increased mineral absorption, or non-specific effects leading to degenerative changes and/or premature death. It is preferable, therefore, to request confirmation of the choice of dose levels by regulatory authorities before running the test. At least three dose levels would be used, with the numbers of rodents in each treatment group being 50 per sex with satellite groups included where necessary. Clinical pathology which is restricted to animals in the chronic toxicity segment would involve the same observations as in sub-acute tests. The control group may be the same size as the test groups or larger by a factor of $\sqrt{n}$ where $n$ is the number of dose levels. Survival of animals to sacrifice should be at least 50% in each group to ensure that data from adequate numbers of animals are available for statistical analysis.

As with sub-acute studies, results of microscopic pathology examinations are crucial. Tumours are diagnosed using internationally accepted guidelines, as is statistical treatment of their organ-specific incidence. Some government agencies, especially in the USA, favour chronic studies including exposure *in utero* in order to investigate the possibility of transplacental carcinogenesis. This requirement may be waived if it can be shown that the test compound and metabolites do not cross the placenta. Weaning animals, usually rodents, are exposed to the test substance and mated within groups. Treatment continues during pregnancy and lactation. After weaning, the offspring are transferred to their parents' diet and treated as in a normal chronic study. Some authorities discourage the use of exposure *in utero* since it is believed that the studies are wasteful of animals, are long and expensive, and difficult to evaluate.[9]

## 2.4. Embryotoxicity and Reproductive Studies

There are numerous possible ways in which a chemical may affect the process of reproduction and/or the development of the foetus once conception has occurred. It is usual to investigate possible effects through the use of two types of test: one, to study effects on reproduction (a multigeneration or fertility study), and the other to study teratogenicity.

The metabolic and pharmacokinetic behaviour of the test substance in pregnant animals may give some insight into the likelihood of reproductive or teratogenic effects, but performing both kinds of study would almost always be necessary.

A typically multigeneration study would be conducted in the rat over at least two filial generations. Observations are made throughout on food intake and bodyweight, and behavioural tests may be performed on the young animals after weaning. Indices are determined for conception, gestation, number and size of litters, birth weight, viability and survival of pups. Thus any effects on male or female fertility and outwardly observable effects on offspring should be detected.

A teratogen is a substance that produces permanent structural or functional damage to the embryo during its development, resulting in congenital malformation or organ malfunction. Teratogens are thought to act during the period of organogenesis, so the administration of test substances in teratology studies is timed to coincide with this. Thus pregnant female rats are dosed by gavage from day 6 to day 15 of gestation. Animals are killed 24–48 h prior to the expected onset of parturition (21 days after conception in the rat) and are examined for the number of mature corpora lutea, the number, weight and sex of mature foetuses, the presence of early resorption sites and of dead or partially resorbed foetuses. The foetuses are weighed, examined visually for abnormalities, and processed to enable both soft tissue and skeletal abnormalities to be detected.

A major problem in extrapolating experimental teratology data to man is that species differ. For example, aspirin is strongly active in the rat and thalidomide is not, whilst the reverse is true in man. In theory, one would perform a study in a species having similarities to man in metabolic profile of the test compound, embryogenesis and placentation. In practice, for food additives, one is restricted to two species: rat and rabbit. Operating experience and detailed knowledge of the nature and incidence of foetal abnormalities is lacking in other species.

## 2.5. Metabolic and Pharmacokinetic Studies[28]

The aim of conducting such studies is to gain a general understanding of the absorption, biotransformation, disposition and elimination of the test substance after both single and repeated doses. One is determining what the test animal, including man, does to the test substance rather than the reverse. Information from metabolic and pharmacokinetic studies should be, and is, used to assist protocol design, species and dose selection, choice of tests *in vitro* and associated activating systems (see Section 2.6). Thus, the toxicological profile of the test substance will be derived using only appropriate procedures and dosages, and the significance of animal data can be extrapolated to man with reasonable confidence.

It is likely that metabolic and pharmacokinetic studies will be undertaken on several occasions during the toxicological programme of an NNS. As soon as isotopically labelled material is available, animal experiments will be of great value. Initially, studies would be conducted in male and female rats with the use (at a later stage) of pregnant animals for examination of placental transfer of the test substance or its metabolites. Studies are often conducted in a second species, for example, the dog.

Assuming that adequate sub-acute and mutagenicity testing (see pages 162 and 165 respectively) had been undertaken, it would be reasonable to conduct some experiments with low doses of non-labelled material in man. Thus a simple balance study combined with some measurements of plasma levels may well be sufficient to establish the similarity of the metabolic and pharmacokinetic profiles in man, rat and dog. However, if there are species differences or metabolic behaviour is complex, more extensive human studies, probably involving the use of [$^{14}$C]-labelled material (ethical considerations permitting[9]), may well be undertaken.

## 2.6. Short-Term Tests

There are many incentives, including the saving of time, money and animal suffering, to develop tests that can lead to a more rapid assessment of the toxicity of a substance. In particular, much effort has been expended over the last decade to develop tests to explore the ability of chemicals to cause mutations and thus alert investigators to the possibility of a mutational and/or carcinogenic hazard if the substance were to be consumed by humans. Tests are performed *in vitro* and *in vivo*, the majority being relatively quick and cheap compared with standard toxicity tests, leading to such descriptions as 'short-term tests' or 'rapid-screening tests'. Fundamental to such tests is the concept of mutation which denotes changes in the genetic material of cells leading to permanent heritable differences in succeeding generations of cells. Many data point to the fact that chemical mutagens interact with cellular DNA at the level of either the individual gene or chromosome. The resulting lesions fall into different categories of gene mutation (for example, 'base-pair' or 'frameshift') or chromosome aberration (for example, structural or numerical).

Mutations in the germ cells of an individual would lead to changes in the genome which may well be expressed in adverse ways in offspring and their descendants, whereas somatic cell mutations have been implicated in the aetiology of several diseases, particularly cancer. Carcinogenic substances showing no evidence of interaction with DNA are called 'epigenetic carcinogens'. Many chemicals have been shown to require conversion by the body to reactive electrophilic intermediates before any DNA binding occurs; such 'metabolic activation' which, in general, results from the body's attempts to convert lipophilic compounds into more polar excretable compounds requires simulation in most assays *in vitro* and is normally achieved by addition of the microsomal supernatant fraction ('S9 mix') of rat liver homogenate.[7]

Several short-term tests give a reasonable correlation of positive results with animal carcinogenicity bioassays, but also give a certain percentage of false positives and false negatives.[29,30] Thus, the concepts of batteries and tiers of tests have evolved in order to reduce the possibility that a genotoxic compound will avoid detection. In the UK, a basic 'package' of four tests has been recommended 'to probe the hereditary machinery sequentially, at various levels of complexity, and take account of the passage of the chemical to the genetic target'.[31] Many other tests are available and may be used to advantage in specific circumstances. The principles and techniques involved in performing the various tests have been well documented elsewhere.[31–7]

Assessing the mutagenic as opposed to the carcinogenic risk in man of a substance through the use of short-term tests is a more difficult process. An Environmental Protection Agency (EPA) discussion document[34] suggests a two-stage 'weight of evidence' approach: first, assessing the genetic toxicity profile of the chemical, and second, determining if the chemical reaches or affects gonadal organs. A series of genetic toxicology tests *in vitro* for gene mutations, and tests *in vivo* for chromosomal effects, are suggested. However, the EPA proposals may be somewhat premature since first, there is no validated economically viable assay *in vivo* for gene mutation, and second, the principle that chemicals *do* cause genetic effects in humans is largely a matter of faith, expressed thus: 'As our knowledge of genetics and disease aetiology increases, it is reasonable to believe that we will become aware of some chemically induced human genetic defects'.[34]

Modern theories of chemical carcinogenesis hold that it is a multi-stage process; the initial damage to DNA being called 'initiation' and the development of the lesion to produce a transformed (malignant) cell-line 'promotion'. The fact that all short-term tests (especially gene mutation tests *in vitro*, such as the Ames test) detect only initiators, and that a plethora of incredibly mutagenic substances have been detected in cooked foods,[39] has led some to question the importance of initiators and to suggest that promoters may play a more crucial role.[40,41] Active research into the many aspects of carcinogenesis may, within the next decade, lead to a reassessment of genetic toxicology tests.

On both sides of the Atlantic, various groups suggest abolition of or reduction in procedures involving the use of experimental animals.[42-4] There are sound scientific reasons as well as emotional and humanitarian considerations for developing tests *in vitro* wherever possible. Although a successful outcome to the various research programmes on the development of non-animal techniques is to be welcomed, it appears most unlikely that all techniques *in vivo* will be superseded in the foreseeable future because of various factors which include:

1.  The complex dynamic and interactive phenomena *in vivo* which cannot be simulated *in vitro*;
2.  The biological and species variation which need to be assessed; and
3.  Pressures from various regulatory authorities, consumer groups and trade unions which encourage an increase in the number and size of animal safety studies.

It is to be hoped that, in the future, the development of techniques *in vitro* will be based on an understanding of toxicological mechanisms and that the amount of live-animal data required for registration of new chemicals can be reduced without prejudice to the interests of the consumer.[45,46]

## 2.7. Safety Evaluation

Safety evaluation is the procedure by which data obtained from various toxicological studies and from other sources (see Table 2) are assessed in order to establish whether the NNS or other substance is safe for human consumption in its particular applications. Ideally the process should be a unique intellectual exercise for each substance, matching the potential hazard with data appropriate in type and quantity.[47] However, the demands by some regulatory authorities and expert committees for all standard tests to be performed in all circumstances has led to the development of 'checklist' safety evaluation[48] which can militate against the development of additives of low toxic potential if funding for toxicological studies is restricted by limited market potential. The application of the checklist approach by some regulatory bodies, differing toxicological philosophies, political and other factors all account for

## TABLE 2
HUMAN RISK ASSESSMENT FOR A TYPICAL NNS: DATA REQUIREMENTS[a]

*Identification and Characterisation*
  1. Name, structure, formula
  2. Specification, impurity profile, analytical procedures
  3. Chemical and physical properties
  4. Method of manufacture and quality-control checks
  5. Storage stability

*Use/Intake Profile*
  1. Quantity employed for all food uses
  2. Food usage pattern, use levels and residue data
  2. Degradation (interaction phenomena in use)
  4. Per capita intake, mean and extreme values
  5. Intake in special subgroups, e.g. children, diabetics
  6. Advantages to consumer

*Toxicological Tests (Species)*
  1. Acute (rat, mouse)
  2. Genetic toxicology
  3. Metabolism and pharmacokinetics (rat, dog, man)
  4. Sub-acute (rat, dog)
  5. Reproductive toxicology, including teratology (rat, rabbit)
  6. Chronic toxicity/carcinogenicity (rat)
  7. Carcinogenicity (mouse)
  8. Special studies, e.g. in biochemistry, immunology, neurotoxicity (various species possibly including man)
  9. Ecotoxicology, biodegradability, environmental impact

*Safety Evaluation*
  1. Significant toxic effects; dose–response; no-effect levels
  2. Extrapolation and relevance of animal data to man
  3. Acceptable daily intake calculations
  4. Identification of special population groups with higher risks

[a] More or less data may be required in specific cases

particular food additives being deemed safe (and therefore available for use) in some countries but not in others.

The predictive value of tests in laboratory animals to estimate hazards to human health relies upon two basic assumptions:

  1.   That there is an appropriate animal model; and
  2.   That, with the exception of some types of chemical carcinogens, it is possible to derive a relationship between the dose of the test compound and the effects observed.

Thus in a review of the toxicological data of a food additive a no-effect level (NEL) in terms of milligrams per kilogram per day or percentage of animal diet per day can be calculated from sub-acute and chronic-toxicity data in the most sensitive species that handles the additive like man. An intake acceptable in human use can be set through application of a safety factor to the NEL assuming that the additive is non-carcinogenic. The purpose of this procedure is to ensure that human intake is well below toxic levels, so the factors used tend to be somewhat arbitrary since what is sought is maximum safety rather than a balance of risk against benefit. It is conventional for many bodies to use a safety factor of 100 (10 to cover extrapolation to man who may be more sensitive than the most sensitive species, and 10 to cover variation of sensitivity of human sub-groups).[49]

The acceptable daily intake (ADI) of an NNS, or any other food ingredient, is calculated by dividing the no-effect level (in $mg\,kg^{-1}\,day^{-1}$) by the appropriate safety factor thus:

$$ADI = \frac{NEL}{Safety\ factor}$$

ADI's are set by JECFA (Joint FAO/WHO Expert Committee on Food Additives) and the concept is used by many (but not all) regulatory authorities. The ADI of a substance provides a regulator with a useful guide to a safe limit of intake; the ADI may be exceeded occasionally but, on average, should not be exceeded.[49]

Because of public concern over the possibility that food additives and other environmental chemicals are a major cause of cancer, the amount of information required by regulatory authorities has increased substantially during the last decade. This has resulted in the development of protocols of increasing complexity and rigidity but which still retain all the elements of empiricism that have been a constant feature of such studies for almost 30 years.

Toxicology is an evolving science and it may be appropriate to adopt a more flexible approach based upon the application of sound scientific principles. A more systematic and logical approach has been proposed by the Food Safety Council. It reduces the element of uncertainty and allows for a more efficient assessment of the potential of a substance for causing injury by developing information in a logical sequence. If adopted, it may be possible to dispense with the arbitrary use of safety factors in calculating what constitutes an acceptable level of intake and may lead to increasing public confidence in the safety of food additives.

There are several factors that can influence the outcome of studies in experimental animals, among which is the selection of appropriate dosages. Because of the frequently expressed need to induce some evidence of toxicity, dosages are employed that are so unrealistic, both with respect to the animals' ability to handle them normally as well as to the anticipated level of human exposure, that it is sometimes impossible to decide whether the responses observed are specific to the test substance or came about as a result of some secondary mechanism. The rat has traditionally been used for most long-term toxicity studies despite its many anatomical and physiological differences from primates. Because of uncertainties in the extrapolation of data from rodents to man, greater confidence in such judgements might be engendered by placing more emphasis on the mechanism of toxic action and less on the acquisition of data of questionable relevance.

Various mathematical models have been developed for extrapolating animal and other data to man for an assessment of risk. For the most part such models[14,50,51] have concentrated on carcinogenesis risk estimation. Unfortunately, owing to a paucity of data, particularly in the low-dose range, estimates of risk can be adduced which differ by several orders of magnitude depending upon which model is used. Until the mechanisms of toxicity, particularly carcinogenesis, are more fully elucidated, it will not be possible to use mathematical extrapolation methods with confidence.

No element of our lives is without risk, and the detectable risks associated with consumption of food additives pale into insignificance when compared with those from other common activities.[50] Thus, risk–benefit analysis has been proposed as a technique for the safety evaluation of NNS's and food additives,[14,50-2] but as yet has not been explicitly accepted by regulatory authorities.

## 3.   SWEETENERS IN CURRENT USE

Brief comments on the toxicological profiles of various substances that are used as intense sweeteners (being considerably sweeter than sucrose) or as bulk sweeteners (being around the same sweetness as sucrose but having a lower caloric contribution) are given below.

### 3.1.   Intense Sweeteners
#### 3.1.1.   Acesulfame-K
There are no published reports of toxicological studies with Acesulfame-K and the following information has been abstracted from the monograph compiled by the Joint FAO/WHO Expert Committee on Food Additives.[3]

Studies in rats, dogs and human volunteers indicate that Acesulfame-K at low dosages is readily absorbed and is excreted unchanged in the urine. Caecal enlargement was reported in rats maintained on diets containing 10 % Acesulfame-K, suggesting that a proportion of the substance was not absorbed. Long-term studies were conducted in rats, mice and dogs but at dietary levels of 0·3, 1·0 and 3 %. No recommendation concerning an ADI was made. Acesulfame-K has been recommended for use in the UK.[5]

### 3.1.2.  Aspartame

The safety of aspartame (N-L-aspartyl-L-phenylalanine methyl ester) and its hydrolysis product, 5-benzyl-3,6-dioxo-2-piperazine, have been extensively investigated in man and several species of laboratory animal. Aspartame is converted in the body to its constituent amino acids while the substituted piperazine derivative is excreted in the urine partly unchanged and partly as phenylacetylglutamine. Studies in adults and children, both normal individuals and those heterozygous for phenylketonuria, have revealed no untoward reactions at dosages of aspartame 10 times greater than the anticipated levels of exposure. This material has had, nevertheless, a chequered regulatory history in the USA. Although approval for its use in a wide range of applications was granted in 1974, objections were raised to its use by children on the grounds that sustained levels of phenylalanine in the plasma might cause brain damage resulting in mental retardation, endocrine dysfunction or both. This particular issue was referred in 1979 to a Public Board of Inquiry, which concluded[62] that the allegations were not supported by the available evidence. A further issue concerned the possible significance of brain tumours found in rats treated continuously with aspartame. Although the Board recommended further testing, the Commissioners concluded that aspartame had been shown to be safe and the initial regulation was reinstated on the approval list in 1981. JECFA at its meeting in 1980, reviewed the data for aspartame and its breakdown product, diketopiperazine, and recommended an ADI of 40 mg kg$^{-1}$ based on a no-effect level of 4000 mg kg$^{-1}$.

### 3.1.3.  Cyclamic Acid

Numerous studies[63] have been conducted to establish the safety of both cyclamic acid (cyclohexylsulphamic acid), which is normally used as either the sodium or calcium salt, and its metabolite, cyclohexylamine, which is produced by the action of intestinal micro-organisms. The extent of conversion varies widely, with a value of 60 % being reported for some individuals. Although cyclamic acid was reported to produce bladder tumours in rats fed a 10:1 mixture with saccharin, subsequent studies do not implicate either cyclamic acid or *cyclo*hexylamine as either potential carcinogens or mutagens. The major toxic effect of *cyclo*hexylamine is that of testicular atrophy but this appears to occur uniquely in the rat. Epidemiological evidence does not suggest an association between the incidence of bladder tumours and the consumption of artificial sweeteners.

A petition to reinstate cyclamate was denied by the Commissioners of the Food and Drug Administration, who asserted that the petitioner had not proved the material to be safe. The UK authorities did not subscribe to this view but, nevertheless, considered that there were sufficient grounds to restrict its use to certain categories, a policy not normally followed with respect to food additives.[5]

### 3.1.4. Saccharin

Synthesised by Remsen and Fahlberg in 1879, saccharin (1,2-benz-isothiazol-3-one-1,1-dioxide) has been used in a wide variety of applications. It is approximately 300 times sweeter than sugar but because of its bitter aftertaste, it is frequently formulated as a 1:10 mixture with sodium or calcium cyclamate.

There is a wealth of data derived from chronic-toxicity studies in several species of laboratory animal, and it is only recently that doubts have arisen concerning its safety—these are based upon reports that saccharin administered at a dietary concentration of $5\%$ to two successive generations of rats resulted in an increased incidence of bladder tumours in males of the second generation.[53]

Apart from the difficulties in the interpretation of studies which involve exposure *in utero*, there is the additional problem that saccharin is not mutagenic and is eliminated from the body unchanged.[1,54]

Saccharin has been the subject of several epidemiological studies which have investigated the possible associations between bladder cancer and the consumption of artificial sweeteners. A Canadian study[55] concluded that there was a $60\%$ increase in the risk of bladder cancer associated with the use of artificial sweeteners in males whereas the risk was reduced to some $40\%$ in females. Several deficiencies were identified in the design of the study and the conclusions have not been substantiated subsequently.[5,56-8] Moreover, a study of approximately 3000 cases of bladder cancer conducted by the National Cancer Institute[64] failed to reveal any association between the consumption of artificial sweeteners, including sodium and calcium cyclamates, and the incidence of the disease. Although it would appear that saccharin is a weak epigenetic carcinogen in rats, it is not known whether a similar mechanism operates in man under normal conditions of exposure.

### 3.1.5. Thaumatin

The sweet principle of *Thaumatococcus daniellii*, a plant indigenous to West Africa, has been associated with a group of basic proteins known collectively as thaumatins. They contain the normal complement of amino acids with the exception of histidine, and apart from their intense sweetness are capable, even when denatured, of extending and enhancing the flavour characteristics of many foods and related products.

Thaumatin is digested as readily as egg-albumen and no untoward effects were observed when it was fed to rats and dogs for 13 weeks at dietary concentrations of up to 8 and 3 %, respectively. The product is devoid of mutagenic activity in bacteria and rodents, and is not teratogenic in rats. Studies in human volunteers indicate that thaumatin is not irritant to the buccal mucosa and does not elicit any allergic response when ingested.[59]

## 3.2. Bulk Sweeteners

### 3.2.1. Hydrogenated Glucose Syrup

A range of products is available in which all free aldehyde groups have been reduced to the corresponding alcohols by hydrogenation. A typical product contains up to 10 % free sorbitol with the remainder comprising hydrogenated di-, tri- and oligosaccharides.[60]

Following ingestion, the di-, tri- and oligosaccharide portion is only partially broken down to sorbitol and glucose. Thus, its laxative effect is less than that of an equivalent amount of free sorbitol. Data from metabolic and sub-acute studies in rat and dog were reviewed by the Committee on Toxicity as part of the Food Additives and Contaminants Committee's review of sweeteners.[5] The material was classified as temporarily acceptable for use in foods on the basis of the present data, and long-term studies were not requested. However, it was recommended that data be collected on intake patterns following its introduction to the market, and on toxicological phenomena common to many of the sugar alcohols, such as altered mineral metabolism and gastrointestinal disturbances.

### 3.2.2. Hydrogenated Palatinose (Isomalt, Palatinit®*)

Isomalt is an equimolar mixture of $\beta$-D-glucopyranosido-1,6-sorbitol and $\beta$-D-glucopyranosido-1,6-mannitol produced by the hydrogenation of 6-$O$-$\beta$-D-glucopyranosyl-D-fructofuranose, which is also known as palatinose,

---

* Palatinit® is a registered trade mark of Süddeutsche Zucker AG, West Germany.

isomaltulose and Lylose®*. Isomalt is about one-half as sweet as sucrose and has been proposed as a substitute for sugar in confectionery products, chewing gum, soft drinks and dessert mixes. There are no published data relating to the toxicity of isomalt, but both constituents have been shown to be slowly hydrolysed by homogenates of human jejunal mucosa with the liberation of glucose, sorbitol and mannitol. Only small quantities of the two disaccharides are excreted in the urine suggesting that they are poorly absorbed.[61]

This conclusion is supported by the absence of a significant increase in either glucose or insulin in the sera of human subjects following the oral administration of 15–100 g isomalt. The product is readily degraded in the large intestine and it has been estimated that approximately 50% of ingested isomalt in man is available for energy production.

### 3.2.3. Mannitol
Mannitol is less widely used than sorbitol. The toxicological profile is similar to that of the other sugar alcohols. Although mannitol is reported to be non-cariogenic, it stimulates the secretion of insulin—a property shared with sorbitol and xylitol—due presumably to the fact that it is partially converted in the body to glucose. The laxative action of mannitol appears to be more pronounced than that of the other sugar alcohols.

### 3.2.4. Sorbitol
The most widely used sugar alcohol, sorbitol, is about half as sweet as sugar. Biochemical studies indicate that it is absorbed only slowly after ingestion but is broken down by intestinal bacteria, which probably explains the laxative effects when excessive amounts are consumed. There are few reports on conventional toxicological studies with sorbitol although various effects, including hyperplasia of the adrenal medulla, were observed following the prolonged administration of sorbitol at a dietary concentration of 20% to rats. Similar effects were observed with xylitol, suggesting a non-specific response. There is additional evidence that sorbitol increases the urinary excretion of essential minerals, in particular, calcium. The mechanism of this response is not known.

### 3.2.5. Xylitol
Xylitol occurs widely in nature and is an intermediate in the metabolism of carbohydrates in man and laboratory animals. It is reported to have the

* Lylose® is a registered trade mark of Tate & Lyle PLC, UK.

same perceived sweetness and calorific value as sugar. Xylitol has been the subject of extensive toxicological investigation and the data reviewed in 1974, and subsequently in 1978, by JECFA. Many of the effects were similar to those obtained with sorbitol. Because of the unrealistically high dosages employed in these studies it is not possible to evaluate with confidence the significance of the effects for man.

### 3.3. Other Sweeteners
There is little or no published toxicological information on other sweeteners mentioned in the literature such as: miraculin, a glycoprotein from *Synsepalum dulcificum* (miracle fruit); monellin, a protein from *Dioscoreophyllum cumminsii* (serendipity berry); β-neohesperidin dihydrochalcone;[2] stevioside,[2] a diterpene glycoside from *Stevia rebaudiana* Bertoni; and trichloro*galacto*sucrose or L-sugar(s).

## 4. CONCLUSIONS

Non-nutritive sweeteners lend themselves to widespread use in many food, paramedical and pharmaceutical products. Consumption by certain groups, for example diabetics, may be high. Additionally, bearing in mind the chequered history of the major NNS's saccharin, cyclamate and aspartame, on the question of safety, the high degree of caution on the part of regulatory authorities in approving any NNS can be readily appreciated. In the face of this situation, however, it is to be hoped that authorities do not fall back on the 'checklist' approach to safety evaluation, nor demand simple black-and-white classifications in circumstances where the problems appear only in various shades of grey.[50]

As a consequence of the number of NNS's either at the food-additive petition stage or in the later stages of development, the 1980s will be a period of active decision-making by regulatory authorities; it is to be hoped that each NNS will be given a *scientific* appraisal on its merits. It would be reasonable to predict that the number of NNS's permitted in various countries around the world will increase, but that the pattern of availability will often vary from one country to another due largely to differing approaches and philosophies on safety evaluation.

### REFERENCES

1. CRANMER, M. F. (1980). *Saccharin—A Report*, American Drug Research Institute, Inc.
2. HIGGINBOTHAM, J. D. Chapter 5 in this publication.

3. International Programme on Chemical Safety (1980). *WHO Additive Series*, No. 16.
4. *Artificial Sweeteners in Food Regulations* (1969). S.I. 1969, No. 1817.
5. *Report on the Review of Sweeteners in Food* (1982). FAC/REP/34, HMSO, London.
6. *OECD Guidelines for Testing of Chemicals, Section 4: Health Effects* (1981). Organisation for Economic Co-operation and Development.
7. *Guidelines for the Safety Assessment of Food Additives* (1980). Report of the Scientific Committee for Food (10th Series), Commission of the European Communities, Luxembourg.
8. *Memorandum on Procedure for Submissions on Food Additives and on Methods of Toxicity Testing* (1965). Ministry of Agriculture, Fisheries and Foods, HMSO, London.
9. *Consultative Document on Guidelines on Toxicity Testing* (1981). Department of Health and Social Security, London.
10. *Procedures for the Testing of Intentional Food Additives to Establish their Safety in Use* (1958). Joint FAO/WHO Expert Committee on Food Additives—Second Report, *Tech. Rep. Ser. WHO*, **144**.
11. FRAZER, A. (1977). In: *Why Additives? The Safety of Foods*, British Nutrition Foundation, Forbes Publications, UK.
12. LECHOWICH, R. V. (1981). *Food Technology* (December), 69.
13. *Evaluating the Safety of Food Chemicals* (1970). National Academy of Sciences.
14. Proposed System for Food Safety Assessment (1978). *Fd Cosmet. Toxicol.*, **16**, Suppl. 2.
15. TRUHAUT, R. (1978). In: *Chemical Toxicology of Food*, C. L. Galli, R. Paoletti and G. Vettorazzi (Eds.), Elsevier/North Holland Biomedical Press, Amsterdam, 3.
16. Principles and Methods for Evaluating the Toxicity of Chemicals, Part I (1978). *Environmental Health Criteria*. World Health Organisation, Geneva.
17. SHIBKO, S. I. and FLAMM, W. G. (1975). *J. Ass. Off. Anal. Chem.*, **58**, 633.
18. JANSSEN, W. F. (1982). *FDA Consumer* (December 1981–January 1982), 6.
19. DAGONI, R. (1981). *Chem. and Eng. News* (March 9th), 26.
20. FOX, J. (1982). *Chem. and Eng. News* (March 8th), 28.
21. CASARETT, L. J. and DOULL, J. (Eds.) (1975). *Toxicology—The Basic Science of Poisons*, Macmillan Publishing Co. Inc., New York.
22. *A Comprehensive Survey of Industry on the Use of Food Chemicals Generally Recognised as Safe (GRAS)* (1972). National Research Council, National Technical Information Service, PB-221, 949.
23. STULTS, V. J., MORGAN, K. J. and ZABIK, M. E. (1982). *School Food Service Research Review*, **6**(1), 20.
24. OSER, B. L. and HALL, R. L. (1977). *Fd Cosmet. Toxicol.*, **15**, 457.
25. ZBINDEN, G. and FLURY-ROVESSI, M. (1981). *Arch. Toxicol.*, **47**, 77.
26. *Notification of New Substances—Consultative Document* (1981). Health and Safety Commission, HMSO, London.
27. FRIEDMAN, L. (1974). In: *Carcinogenesis Testing of Chemicals*, L. Golberg (Ed.), CRC Press, Cleveland, Ohio, 21.

28. ROE, F. J. C. (Ed.) (1970). *Metabolic Aspects of Food Safety*, Blackwell Scientific Publications, London.
29. GOLBERG, L. and DANIEL, J. W. (1980). In: *The Principles and Methods in Modern Toxicology*, C. L. Galli, S. D. Murphy and R. Paoletti (Eds.), Elsevier/North Holland Biomedical Press, Amsterdam, 341.
30. WILLIAMS, G. M. and WEISBURGER, J. H. (1981). *Ann. Rev. Pharmacol. Toxicol.*, **21**, 393.
31. *Guidelines for the Testing of Chemicals for Mutagenicity* (1981). Department of Health and Social Security, Report on Health and Social Subjects, **24**, HMSO, London.
32. *The Testing of Chemicals for Carcinogenicity, Mutagenicity and Teratogenicity* (1975). Department of Health and Welfare, Canada.
33. GRIESEMER, R. A. and DUNKEL, V. C. (1980). *J. Environmental Path. and Toxicol.*, 565.
34. ANON. (1980). *Fd Cosmet. Toxicol.*, **18**, 683.
35. CASCIANO, D. A. (1982). *Food Technology* (March), 48.
36. TAYLOR, S. L. (1982). *Food Technology* (March), 65.
37. RÖHRBORN, G. (1980). *Arch. Toxicol.* (Suppl. 4), 3.
38. Mutagenicity Risk Assessments: Proposed Guidelines, Environmental Protection Agency (1980). *Federal Register*, **45**, November 13th, 74984.
39. PARIZA, M. W. (1982). *Food Technology* (March), 53.
40. GOLBERG, L. (1982). *Chem. and Ind.*, 354.
41. SLAGA, T. J. (1981). In: *Nutrition and Cancer: Etiology and Treatment*, G. R. Newell and W. M. Ellison (Eds.), Raven Press, New York.
42. LANE-PETTER, W., FELL, H. B. and MELLANBY, K. (1977). *Biologist*, **24**(5), 229.
43. HOLDEN, C. (1982). *Science*, **215**, 35.
44. PATON, W. D. M. (1978). *Biologist*, **25**(4), 142.
45. PAGET, G. E. (1978). *Biologist*, **25**(4), 145.
46. ANON. (1978). *Nature*, **274**, 413.
47. GOLBERG, L. (1975). *J. Ass. Off. Anal. Chem.*, **58**, 635.
48. SHARRATT, M. (1977). In: *The Evaluation of Toxicological Data for the Protection of Public Health*, Commission of the European Communities.
49. *Food Additives and the Consumer* (1980). Commission of the European Communities.
50. DOULL, J. (1981). In: *Food Safety*, H. R. Roberts (Ed.), J. Wiley and Sons, London.
51. *A Proposed Food Safety Evaluation Process* (1982). Food Safety Council.
52. COHEN, B. L. (1978). *Nature*, **274**, 492.
53. ARNOLD, D. L. *et al.* (1980). *Toxicol. Appl. Pharmacol.*, **52**, 113.
54. BALL, L. M., RENWICK, A. G. and WILLIAMS, R. T. (1977). *Xenobiotica*, **7**, 189.
55. HOWE, G. R. *et al.* (1977). *Lancet*, **2**, 578.
56. KESSLER, I. and CLARK, J. P. (1978). *J. Amer. Med. Assoc.*, **240**, 349.
57. MORRISON, A. S. and BURING, E. (1980). *N. Engl. J. Med.*, **302**, 537.
58. WYNDER, E. L. and STELLMAN, S. D. (1980). *Science*, **207**, 1214.
59. HIGGINBOTHAM, J. D., SNODIN, D. J. and DANIEL, J. W. (1982). *The Toxicologist*, **2**(1) (Abstract No. 614), 176.

60. DOSON, A. G. and WRIGHT, S. J. C. (1982). *Food Flavourings, Ingredients, Packaging and Processing* (September), 29.
61. GRUPP, U. and SIEBERT, G. (1978). *Res. Exp. Med. (Berl.)*, **173**, 261.
62. ANON. (1981). *Food Chem. News*, July 20th, 54; *Federal Register*, **46**(142), July 24th, 38284.
63. ANON. (1980). *Federal Register*, **45**, September 16th, 61474.
64. HOOVER, R. N. and STRASSER, P. H. (1980). *Lancet*, **1**, 837.

*Chapter 7*

# THE FATE OF NON-NUTRITIVE SWEETENERS IN THE BODY

A. G. Renwick

*Faculty of Medicine, University of Southampton, Southampton, UK*

*SUMMARY*

*Cyclamate is converted to a metabolite, cyclohexylamine, which is more toxic than the parent compound so that the acceptable daily intake (ADI) for cyclamate will be determined by its metabolism. Saccharin is a strong organic acid which increases the incidence of bladder tumours when fed to rats at very high dietary levels. Its metabolism, or lack thereof, is a key issue in its classification as either a genetic or epigenetic carcinogen, and may affect profoundly how it is regulated. Aspartame is metabolised to amino acids and methanol; its potential toxicity is related to the possible consequences of high blood levels of aspartic acid, phenylalanine and methanol.*

## 1. INTRODUCTION

In this chapter the fate of two non-nutritive sweeteners, cyclamate and saccharin, and a new low-calorie sweetener, aspartame (Fig. 1) will be considered with particular emphasis on their metabolism and excretion since these have a major bearing on their safety (in use).

All three compounds either have received or are about to receive wide usage in human food supplies and, as 'non-essential' food additives, have been the subject of considerable debate both in the scientific and lay press.

179

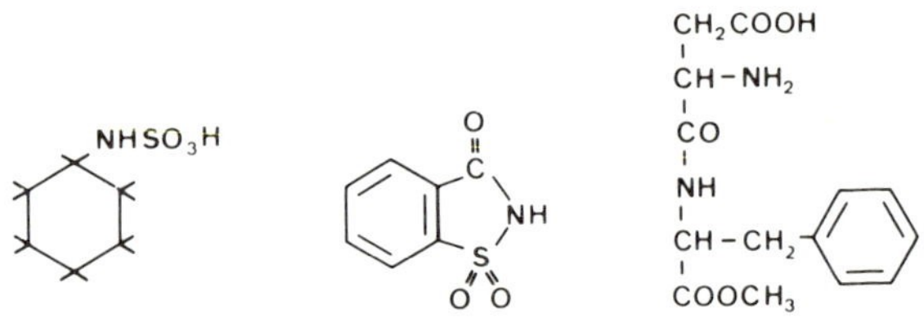

Fig. 1.  Non-nutritive and low-calorie sweeteners.

Saccharin has been used since the turn of the century in most countries but, at the time of writing, is under a cloud of suspicion; cyclamate was used from the mid 1950s until 1969 when it was banned in some, but not all, countries, and it may well be reintroduced pending a re-examination of the evidence; aspartame is about to enter the lists, having received Food and Drug Administration (FDA) approval recently for use as a food additive, and it will doubtless be the source of future debate.

Although the chronological order of their discovery and use was saccharin, cyclamate, aspartame, it was the banning of cyclamate in 1969 which stimulated a major re-examination of saccharin. Thus the compounds will be discussed here in the order cyclamate, saccharin, aspartame.

## 2.  CYCLAMATE

Cyclamate (cyclohexylsulphamate) was discovered in 1944.[1] Early studies on its fate suggested that it was poorly absorbed from the gut and excreted unchanged.[2] However, with the development of more sophisticated analytical techniques, it became apparent that cyclamate underwent some biotransformation in the body. In 1969, it was banned as a food additive in the USA and soon after in the UK and various other countries[3]—a decision that remains controversial to this day. The basis for the ban was a study[4] in which rats were fed cyclamate and saccharin in a 10:1 ratio at high doses (up to 5 % of the diet or $2 \cdot 5\,\mathrm{g\,kg^{-1}\,day^{-1}}$) throughout their lifetime. During the study it was realised that cyclamate could be metabolised to cyclohexylamine and this compound was added (at a dose equivalent to about 10 % metabolism of the cyclamate component) to the diet of some of the rats. After 2 years an increased incidence of bladder tumours was found in male rats fed the highest dietary level. On the basis of this experiment it i impossible to conclude which agent was responsible, a fact acknowledged by the authors when the full paper was published.[5] However, in the political climate of the late 1960s a regulatory conclusion was required, and the long use of saccharin, together with the finding of a bladder tumour in one male rat in a group given cyclohexylamine alone,[4] tipped the balance against cyclamate. Despite the paucity of data confirming this conclusion, it has since become apparent that the metabolism of cyclamate is an essential element in its toxicological evaluation since the metabolite, cyclohexylamine, is considerably more toxic than cyclamate.

The fate of cyclamate in the body is dominated by the presence of the strongly acidic sulphamic acid group ($pK_a$ about 2).

## 2.1.  Absorption

Early studies[6,7] in the rat using radiolabelled cyclamate showed that 30–60 % of an oral dose was recovered in the faeces, whereas parenteral administration gave much lower faecal elimination (2–20 %). This suggests that cyclamate is incompletely absorbed from the gut. Extensive excretion of an oral dose of cyclamate in the faeces has been shown in rabbits,[8] guinea-pigs,[8] pigs[9] and monkeys.[10] This must be due to poor absorption because cyclamate undergoes negligible excretion in the bile after either oral[11] or parenteral[8] dosing to rats. Studies in the rabbit[12] suggested that the absorption rate was enhanced slightly by the co-administration of caffeine or citric acid (which are ingredients of cyclamate-containing beverages). These conclusions assume that the rate of absorption exceeds that of excretion, but in view of subsequent findings with saccharin (see Section 3.1) which is absorbed more extensively than cyclamate,[13] they must be viewed with some caution. Related studies using rat intestinal preparations in situ[13] have demonstrated that the absorption of cyclamate was enhanced by caffeine and citric acid.

In man, the extent of absorption of an oral dose is variable and the total percentage recovered in the faeces has been reported as 65,[14] 18–36,[15] 25–51,[16] and 33–60 %.[8] The absorption of cyclamate is slow and the time to peak plasma concentrations is 6–8 h.[16] The plasma half-life after oral dosing has been reported to be 8 h[16] but this value is probably a reflection of slow absorption because the elimination half-life after intravenous dosing is 1–2 h (70–90 % of the dose was recovered in the urine within 3 h).[14] Similarly, in the rat[9] urinary elimination after oral administration was slower than after subcutaneous injection. A comparable situation arises with saccharin, which has been studied more intensively and is discussed in greater detail in Section 3.1.

## 2.2. Distribution

Cyclamate binds to circulating plasma proteins[16,17] from which it may be displaced by highly bound drugs such as phenylbutazone.[18] The concentrations of [$^{35}$S]cyclamate in rat tissues 2·5 h after an intravenous dose were (with the exception of the liver) lower than the blood level;[6] however, at later times the levels in most tissues were similar to the blood concentrations while the kidney usually contained the highest concentration. The total body load represented less than 1 % of a single dose at 24 h[7,19] and less than 0·1 % at 72 h.[7] Similar results were reported for the rabbit[6] and the dog.[6,7] Also, it was found that the sweetener, but not cyclohexylamine, diffused to only a limited extent across the primate placenta.[20] In rats, a test dose of [$^{14}$C]cyclamate after a period of chronic administration showed a similar distribution to a single dose and the highest steady-state tissue concentrations were in the kidney.[19] The kidney showed the highest concentration of the metabolite cyclohexylamine (see Section 2.3), but it was also present in most other tissues; however, the authors did not report the concentration in the testes which are the main site of cyclohexylamine toxicity.

## 2.3. Metabolism

At the time of its introduction as a food additive, cyclamate was believed to be eliminated unchanged in the urine and faeces.[2] In 1966, cyclohexylamine was detected in the urine of the dog and man after oral cyclamate administration and this report[21] led to a number of studies aimed at defining the extent, site and significance of this apparent metabolic pathway (Fig. 2). No cyclohexylamine was detected in the urine of 100 volunteers on strict cyclamate-free diets[9] so that any detected after cyclamate administration must be a metabolite of the sweetener.

Since this initial report, cyclohexylamine has been detected in the urine of rats, rabbits, guinea-pigs, dogs, pigs, monkeys and humans given cyclamate orally. The amounts of cyclohexylamine detected in the urine by

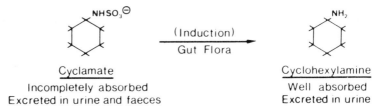

Cyclamate
Incompletely absorbed
Excreted in urine and faeces

Cyclohexylamine
Well absorbed
Excreted in urine

FIG. 2.   The metabolism of cyclamate.

various workers show a very wide range of values (0–60 % of the dose of cyclamate) between different individuals and species studied.

## 2.3.1. The Induction of Cyclamate Metabolism

The main differences between early studies in which cyclamate was reported to be unmetabolised[6,7] and subsequent investigations in which metabolic change was detected (see below for references) were the development of specific assay procedures for cyclohexylamine and the exposure of animals and humans to diets containing cyclamate prior to the metabolism study. Thus, [14C]cyclohexylamine was detected in the urine of rats, rabbits and guinea-pigs that had been fed cyclamate chronically in the diet prior to a dose of [14C]cyclamate, but not in the same animals before chronic administration.[8]

The time taken for the induction of cyclamate metabolism on continued administration shows large differences between animal species and among different individuals. Some workers have reported that rats given diets containing cyclamate were able to metabolise up to 45 % of a test dose to cyclohexylamine[8,11,19,22–5] whereas others[9] did not produce 'converter rats' by chronic administration. The unpredictability of this induction process in rats is shown in studies in which rats were given 0·5 % cyclamate in the drinking water *ad libitum*. In one study,[19] five groups of rats did not develop the ability to metabolise cyclamate (> 1 %) until 6, 11, 16, 26 and 31 weeks of treatment, but most rats in each group became converters at approximately the same time. A total of 24 out of 26 rats became converters within seven months. The author has performed two studies on the metabolism of cyclamate in rats during chronic ingestion, which were separated by a period of about five years.[8,26] In the initial study, extensive converters (up to 40 % of the dose) were observed after three months' treatment. During the second study using groups of rats from three different suppliers there was a very gradual development of cyclamate → cyclohexylamine metabolism, with an average of only about 1 % metabolism after twelve months treatment. The time taken for individual rats to become good converters (> 1 % metabolism) ranged from eight to twenty-three months, after which they either maintained their acquired ability or showed variability. Some individuals remained relatively poor converters (< 0·5 % metabolism) for up to two years whilst sharing cages with good converters. This is in contrast to other studies which reported that the introduction of a converter rat into a cage of non-converters brought about a transfer of metabolising ability (presumably by coprophagy; see below).[9,24] Other variables that may assist the development of cyclamate metabolism include keeping rats in wire-mesh cages rather than over wood shavings[26] (which probably results in increased coprophagy) and starting chronic cyclamate administration during development of the gut flora at weaning.[26]

The time taken for the development of cyclamate metabolism in other animal species has been less well defined but studies suggest that guinea-pigs take about eight days,[27] rabbits about one month[28] and pigs between one and three weeks[9] of continuous dietary intake of cyclamate to induce a significant degree of metabolism.

As might be predicted by these animal studies, the metabolism of cyclamate in man is also subject to a period of induction. Although cyclohexylamine has been detected in the urine of human subjects given a single dose of cyclamate[8,29-32] it was usually at a low level ($< 1\%$ of the dose) and excreted in the second to fourth day after dosing,[8,29] whereas most unchanged cyclamate was eliminated within the first two days after oral administration.[8,14,29] In addition, it should be realised that most of these studies where performed when cyclamate was available for use in the diet, and there may have been some degree of induction before the study. During regular daily administration of cyclamate the percentage of the dose excreted as cyclohexylamine in the urine increases until it reaches a plateau. In subjects taking 0·5–3 g per day the time taken to reach a plateau value has been reported as 4 days,[9] 10 days,[33] 3–5 days,[15] between 5 and 10 days[8] and 9 days.[30] After the establishment of a plateau for a particular individual the extent of metabolism is subject to large temporal variations with the maximum in any 24-h period being up to twice the average plateau value. Continued administration is important for the maintenance of metabolic activity since removal of cyclamate from the diet results in rapid loss of the induced ability to metabolise cyclamate in both rat[8,19] and man.[8]

### 2.3.2. The Occurrence and Extent of Cyclamate Metabolism in Man

The large individual differences in the extent of metabolism in rats,[8,11,19,22,23,26,28] rabbits,[8,28] guinea-pigs[8,27] and pigs[9] given cyclamate continuously are equally apparent in man.

Several studies have attempted to define the proportion of the human population able to metabolise significant amounts of cyclamate, by giving a large number of subjects a single oral dose of cyclamate and measuring urinary cyclohexylamine for up to 24 h after. However, because of the nature of the induction process, such studies are of limited value and tend to underestimate the incidence of metabolism. In addition, some of the published studies were only qualitative and no detailed methodology was described to allow an assessment of sensitivity. Based on such studies, the incidence of significant metabolism ( $>0.2\%$ of the dose where quantified) has been reported as $8\%$ (37 subjects[34]), $11\%$ (156 subjects[30]), $12\%$ (100 subjects[32]), $14\%$ (69 subjects[35]) and $29\%$ (118 subjects[31]) or an average of $16\%$ of the population (76 out of 480 subjects investigated).

Of greater reliability are those studies in which 24-h urine samples, produced after at least three consecutive daily doses of cyclamate, were analysed for cyclohexylamine. Such studies have reported that the incidence of conversion in man is $11\%$ (35 subjects[33]), $20\%$ (5 subjects[36]), $26\%$ (141 subjects[9]) or $33\%$ (3 subjects[8]) in Europe, and $42\%$ (24 subjects[37]) or $64\%$ (11 subjects[15]) in North America. Thus the average incidence is about $27\%$ of the population (60 out of 219 subjects studied).

Significant information lacking in these data is the incidence of cyclamate metabolism in the general population when exposure to cyclamate was via its use as a food additive. Analysis of urine samples collected routinely from 139 psychiatric patients[34] revealed that $37\%$ of known cyclamate users excreted cyclohexylamine, although quantitative data were not given. Such information was provided by Asahina *et al.*[29] for the 24-h urinary excretion of both cyclamate and cyclohexylamine in 50 Japanese subjects taking a normal diet during the summer of 1969. All urine samples contained cyclamate (0·9–744 mg) and most contained cyclohexylamine (0–129 mg). The daily dose of cyclamate was not measured but can be estimated approximately from the urinary cyclamate plus cyclohexylamine, assuming that the faeces contained an amount of cyclamate equal to that in the urine. Using these values for the 35 subjects who excreted clearly measurable amounts of cyclamate in the urine ( $>10$ mg), the range of metabolism was 0–67% of the daily intake, and $60\%$ of the population excreted $1\%$ or more as cyclohexylamine. Extensive metabolism ( $>5\%$ ) of cyclamate occurred in $34\%$ of the study group, but only $4\%$ of the group excreted more than $20\%$ as cyclohexylamine. However, it must be emphasised that such calculations assume that the individuals had a constant regular intake. Since most cyclamate is excreted on the first day of dosing whilst cyclohexylamine is formed and eliminated 24 h later, the apparent percentage conversion calculated above would be influenced profoundly if intake were intermittent during the three days preceding the study.

Thus on the basis of these studies it appears that 20–30% of populations on a Western diet acquire the ability to convert cyclamate to cyclo-hexylamine on continual dietary intake. However, most of these subjects will excrete low levels of cyclohexylamine, and very extensive metabolisers ($>20\%$ of the dose) comprise only a small proportion of the population. Figure 3 gives the population distribution for cyclamate metabolism based on all subjects given regular doses of cyclamate and for whom 24-h cyclohexylamine excretion was quantified.[8,9,15,30,33,35,36] However, it should be realised that in two of the studies[30,35] (representing 18 subjects in

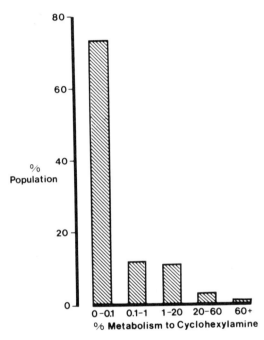

FIG. 3. The distribution of cyclamate-metabolising ability in the human population during controlled chronic intake. The values are derived from all available published data and represent a total of 221 individuals given from 0·5 to 5 g of cyclamate daily for at least three days.

Fig. 3) the individuals were selected (from 121 studied) for their ability to excrete traces of cyclohexylamine after a single dose of cyclamate. Thus the values given in Fig. 3 probably overestimate the proportion of 'good' metabolisers in a normal population; yet despite this, only about $4\%$ of subjects excreted more than $20\%$ of the cyclamate as cyclohexylamine during a period of carefully controlled, consistent intake.

*2.3.3. The Site Metabolism of Cyclamate to Cyclohexylamine*
Nearly all animal studies have indicated that after continued intake of cyclamate, its conversion to cyclohexylamine is brought about exclusively by the gut microflora. Using tissues from animals known to metabolise cyclamate *in vivo*, the formation of cyclohexylamine *in vitro* has been demonstrated in the gut contents or faeces of rats,[19,24−6,38] guinea-pigs,[27,38,39] rabbits[38] and dogs,[40] but no metabolism has been detected with homogenates of other tissues of rats,[38] guinea-pigs[27,38] or rabbits[38] or in the isolated perfused rat liver.[23]

The importance of the gut flora *in vivo* has been shown by the complete suppression of cyclamate metabolism by oral antibiotics in rats[19,41] and pigs,[9] and by the negligible metabolism of cyclamate after intravenous[41] or intraperitoneal[8,19] injection in chronically treated rats and after subcutaneous injection in pigs.[9] Kojima and co-workers claimed, on the basis of incubation results, that the liver was the site of cyclamate metabolism in the rat[42] and subsequently[28] that the liver was the site of metabolism in the rabbit but not in the rat. However, these studies are open to criticism since only trace amounts of cyclohexylamine were present on incubation of cyclamate with rat liver[42] and it was not clear whether this and the metabolites of cyclohexylamine detected[28] (see below) arose from liver metabolism or from the period of pre-treatment with cyclamate orally. Studies using antibiotics[43] suggested that the gut flora were the site of metabolism in the rat but not in the rabbit, but the levels of cyclohexylamine were very low, indicating that the animals had not been induced to metabolise significant amounts of cyclamate.

Evidence for the involvement of the gut flora in cyclamate metabolism in man must of necessity be circumstantial. Early studies[15,33] reported that the faeces of induced subjects were unable to metabolise cyclamate *in vitro*. A subsequent paper[38] showed that the faeces from such individuals could produce [$^{14}$C]cyclohexylamine on incubation with small amounts of [$^{14}$C]cyclamate, but that prior incubation in a medium containing 0·5 % cyclamate resulted in a large (up to 300-fold) reduction in the cyclamate-metabolising capacity of the faecal organisms. This probably explains the absence of conversion reported previously, since in one study[15] the concentration of substrate used was 1·0 %. Evidence for the importance of the gut flora *in vivo* in man includes the finding that metabolism of cyclamate by faeces *in vitro* mirrored the induction and loss of metabolism *in vivo* during and after a period of continuous cyclamate administration,[38] and the complete suppression of cyclamate metabolism *in vivo* by oral antibiotics.[9]

Various organisms able to metabolise cyclamate to cyclohexylamine
have been isolated from the gut contents or faeces of human and animal
converters. Clostridia have been isolated from dogs[40] and rats,[38]
*Pseudomonas* and *Corynebacterium* from guinea-pigs,[39] enterobacteria
and clostridia from rabbits[38] and enterococci from man.[38] It is apparent
that no one organism is responsible for cyclamate metabolism in all animal
species, and indeed a variety of organisms may be responsible in different
individuals of the same animal species. For example, clostridia were the
only active organisms isolated from converter rats[38] and their numbers
increased during two experiments in the same laboratories.[8,26] However,
another almost identical study[19] reported no increase in clostridia whilst
the conversion *in vivo* was suppressed by neomycin[19,41] which is not active
against clostridia. In addition, although enterococci[38] were the only
organisms isolated from one human subject's faeces that were able to form
cyclohexylamine, other human converters with different diets may have
other cyclamate-metabolising organisms present in their intestinal tracts.

## 2.3.4. Other Metabolites of Cyclamate

Dicyclohexylamine has been reported in the urine of converter rats given
[14C]cyclamate[23] but the methods of analysis and characterisation were
inadequate and the possibility of dicyclohexyl sulphamate in the dose was
not investigated. This finding was not substantiated.[16] Cyclohexylhydroxylamine (*N*-hydroxy-cyclohexylamine) was reported in the urine of
cyclamate-treated monkeys[10] and man[40] but no methods of characterisation were given. Cyclohexanol and cyclohexanone have also been
reported as trace metabolites of cyclamate in rat and rabbit,[28,42,43] guinea-
pig,[27] monkey[10] and man.[40] These probably arose from the metabolism of
cyclohexylamine since their appearance in the urine was generally
coincidental with this metabolite. The sulphate formed from cyclamate was
utilised and incorporated into bacterial protein and converted to hydrogen
sulphide.[25]

## 2.3.5. The Fate of Cyclohexylamine

A study in 1937[44] suggested that cyclohexylamine given to dogs was
destroyed completely and that none appeared in the urine. This gave rise to
suggestions[24,45] that measurement of urinary cyclohexylamine after
cyclamate ingestion may represent merely the 'tip of the iceberg' and that
considerably more cyclamate was metabolised than could be recovered as
cyclohexylamine in the urine. This was refuted by studies on the fate of
[14C]cyclamate in converter rats in which cyclamate and cyclohexylamine

accounted for all[11,19] or most[8] of the urinary $^{14}C$, and no $^{14}CO_2$ was found in the expired air.[8]

A preliminary report[46] on the fate of [$^{14}C$]cyclohexylamine in the rabbit showed that 68% was recovered in the urine in 60 h—45% as unchanged cyclohexylamine, 0·2% as cyclohexylhydroxylamine in a conjugated form and 2·5% as cyclohexanone oxime (an artefact). Chromatography of the urine separated a glucuronide and two aliphatic amino compounds, but the minor of these corresponded to cyclohexylamine, and this discrepancy makes interpretation difficult. A study in which rats, dogs and man were given [$^{14}C$]cyclohexylamine[47] showed that most $^{14}C$ was recovered in the urine unchanged and that metabolites accounted for 11–27% in animals and about 2% in man.

A more complete study in rat, rabbit, guinea-pig and man[48] showed that the urine was the major route of elimination for $^{14}C$ after [$^{14}C$]cyclo-hexylamine, and accounted for about 90% of the dose in all species. In rats, guinea-pigs and man about 90–95% of the urinary $^{14}C$ was the parent compound while in rabbits only about 45% was unchanged. In rats the main route of metabolism (Fig. 4) was ring hydroxylation to various aminocyclohexanols, whereas in rabbits and guinea-pigs ring hy-droxylation and deamination to cyclohexanol and trans-cyclohexane-1,2-diol occurred. In man only the deaminated products were detected[48] and these were excreted as conjugates. This agrees with the data of Hengstmann et al.[31] who recovered more than 95% of cyclohexylamine unchanged in the urine of man with no detectable cyclohexanol, cyclohexanone or cyclohexylhydroxylamine. Cyclohexylhydroxylamine and cyclohexanone were detected as trace metabolites in rabbit urine only.[48] The latter compound is the initial product of deamination of cyclohexylamine by rabbit hepatic microsomes,[49] and subsequently undergoes reduction to cyclohexanol.

Cyclohexylamine is distributed extensively to the tissues and has an apparent volume of distribution of about 3 litres $kg^{-1}$ meaning that little of the compound remains in the plasma, and an elimination half-life of 2–4 h in rat, dog and man.[31,47] Thus only slight differences exist in the route and extent of cyclohexylamine metabolism and excretion, and in the rat and man at least 90% of any cyclohexylamine produced from cyclamate should be measurable as such in the urine.

## 2.3.6. *Other Sulphamates as Non-Nutritive Sweeteners*
Since cyclamate (cyclohexylsulphamate) has run into toxicological prob-lems, some interest has been shown in the possibility of using other

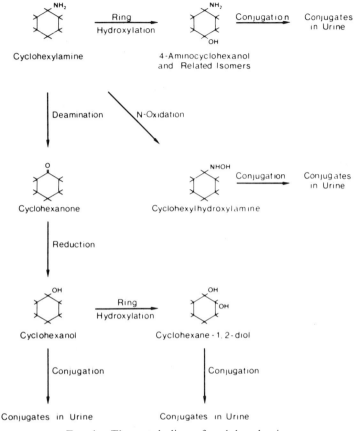

FIG. 4.    The metabolism of cyclohexylamine.

alicyclic sulphamates as safer alternatives. However, for any such compounds three key interrelated issues have to be resolved:

1.    What is the potential toxicity of the amine?
2.    Is the sulphamate a substrate for microbial enzymes? and
3.    Is the sulphamate able to induce its own metabolism?

Studies using an enzyme isolated from *Pseudomonas* sp.[50] or a rat faecal preparation[26] have indicated that the 'sulphamatases' induced by chronic cyclamate administration in guinea-pigs and rats, respectively, can hydrolyse a range of related substrates. Studies on the metabolism of 3-methylcyclopentylsulphamate *in vivo* in cyclamate-metabolising rats support this proposition.[26] Studies *in vivo* on the metabolism of a range of sulphamate substrates in rats[51–3] showed negligible metabolism ( < 0·5 %) but these results are of limited predictive value since the maximum period of continuous intake was only nine days.

No studies have attempted to define the extent of induction of metabolism by long-term chronic administration of any of the alternative sulphamates. Such data are essential before any valid conclusion can be reached on these agents.

## 2.4. Excretion

As discussed above, the absorption of cyclamate from the gut is slow, and this limits the rate of elimination after an oral dose, but if cyclamate is given by injection it is eliminated rapidly[9,14] and recovered in the urine. Although the renal elimination of cyclamate has received scant attention compared with that of saccharin (see Section 3.4.1), in the rat cyclamate also undergoes active secretion by the renal tubule at a rate which exceeds the clearance of inulin.[54] However, attempts to demonstrate competition between cyclamate and *p*-aminohippurate (PAH) were unsuccessful, possibly due to the use of a dose of cyclamate which was one-tenth of that which saturated elimination capacity.[54] The bile has negligible importance in the elimination of absorbed cyclamate.[8,11]

## 3.  SACCHARIN

Saccharin which has been used as a non-nutritive sweetener for most of this century, has recently been implicated as a possible carcinogen to the urinary bladder. The evidence for this is an increased incidence of bladder tumours in experimental animals under the extreme conditions of high dietary levels (5 % or more) and exposure over two generations. However, extensive epidemiological studies have produced little or no supportive evidence in man. Thus in the following discussion certain key issues will be emphasised:

1.   Does the anionic saccharin molecule undergo metabolism to an electrophilic species capable of binding to DNA?
2.   Does the urinary bladder accumulate saccharin during chronic intake? and
3.   Is the rat a valid model for man in respect of the fate of the sweetener in the body?

Saccharin is a strong organic acid, like cyclamate, and shares certain characteristics, such as slow absorption, no measurable metabolism by the tissues and rapid renal elimination. The chemical reactivity of the molecule is dominated by the labile, acidic hydrogen atom, which ionises in aqueous solution but may be substituted by acyl or alkyl radicals under appropriate anhydrous conditions. Addition reactions to the 3-carbonyl group have been reported.[55] The isothiazole ring undergoes hydrolytic cleavage at high temperature in the presence of acid and alkali to yield 2-sulphobenzoic acid and 2-sulphamoylbenzoic acid, respectively (see Fig. 7, p. 196). However, in aqueous solution at pH 7 it undergoes only 0·3 % hydrolysis after heating at 100 °C for 1 h, and negligible hydrolysis at pH 8.[56]

### 3.1. Absorption

Saccharin is absorbed incompletely and to a variable extent from the gut after oral administration. The proportion of the dose recovered in faeces is 3–30 % in rats,[57–63] 2–4 % in guinea-pigs,[58] 4–11 % in rabbits,[62] 2–4 % in monkeys[64] and 1–9 % in man,[62,65,66] the remainder being detected in the urine. The material present in faeces is almost certainly unabsorbed saccharin since little is found in faeces after an intravenous dose in either rat[67] or man,[66] and negligible amounts of this low-molecular-weight acid are excreted in the bile of rats.[59,62] There is an indication in the literature of a decrease in the extent of absorption at high doses,[61,63,67] but, because of the wide range of values found in different studies, this may not be of great importance. A further variable that influences the absorption of saccharin is whether or not the animals were fasted prior to dosing. Although the total recovered in urine during 24 h is not greatly affected, the initial concentration of saccharin in blood[60] or plasma[67] is higher in fasted than in fed rats. In rats the concentration of saccharin in plasma after an oral dose[60,67] shows an early peak, about 15 min after dosing, followed by a variable slow decline. However, the concentration is about 30-fold less than after a similar dose given by intravenous injection[67] (Fig. 5). Thus it appears that the plasma concentration–time curve after oral administration is determined by the absorption rate from the gut, with an early peak due to more rapid absorption from the stomach and possibly the upper intestine, which may be altered by the presence of food, followed by very slow absorption from the lower intestine. Similarly, route of administration differences in pharmacokinetics and effects of food were found in man[66] (Fig. 6).

### 3.2. Distribution

Once saccharin enters the blood, it binds reversibly to plasma protein.[68] The extent of this binding has been reported as 69–86 %,[67] 24–35 %[69] and

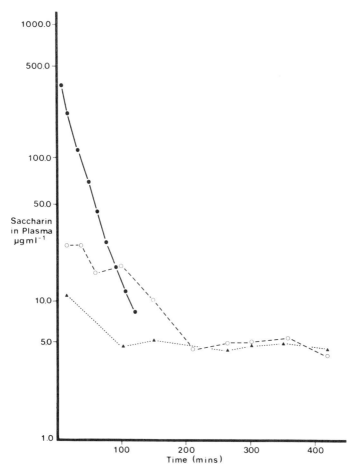

FIG. 5. The concentrations of saccharin in the plasma of rats given $0.1 \, g \, kg^{-1}$ orally and as an intravenous bolus dose (data adapted from Reference 67). ●, Intravenous; ○, oral-fasted; ▲, oral-fed.

$3\%$[70] in the rat, and $75–80\%$ in man.[66,71] The very low value reported in the rat[70] is probably an artefact since the method of analysis (Sephadex chromatography) would allow dissociation of the sweetener–protein complex. The compound distributes rapidly to the tissues after intravenous dosing.[67] The tissue distribution after a single oral dose in rats[60-2] shows the highest concentrations in well-perfused tissues such as the liver and lungs, and in the organs of elimination, i.e. the kidneys and urinary

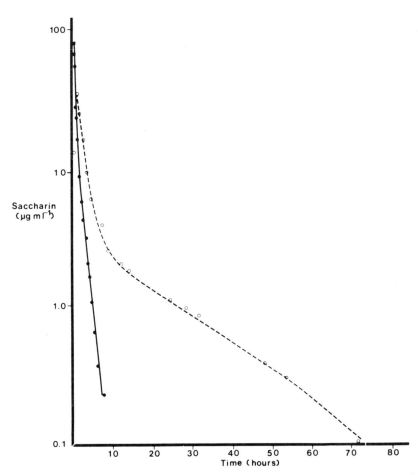

FIG. 6. Plasma concentration–time curve after oral and intravenous doses of saccharin in a human volunteer. ○, Oral dosage (2 g); ●, intravenous dosage (0·775 g) (from Reference 66).

bladder. The concentration in the bladder wall[60,61] was higher than the corresponding value for blood, but highly variable values were reported, and the ratio of saccharin in bladder:blood during the first 24 h after dosing was between 0·6 and 31. A high variability was also found in the kidney (0·3–25). This variability probably reflects problems of contamination of both tissues by urine, because the concentration in urine after a single intravenous dose is about 100 times that in the bladder wall.[67] Studies of tissue levels after multiple oral doses[60] led to the suggestion, adopted in the National Academy of Sciences review of saccharin,[72] that saccharin accumulated on repeated intake, particularly in the urinary bladder. However, in a study on the steady-state concentrations in rats fed 1–10 % saccharin-containing diets[67] it was found that the concentration in the bladder, although higher than other tissues apart from the kidney, was only about 1·6 times the plasma level. This result indicates that there is no excessive accumulation in the bladder.

In cancer bioassays an increased incidence of bladder tumours is associated with second-generation ($F_1$) male rats fed 5 % or more saccharin in the diet. Since saccharin is not metabolised (see Section 3.3), this effect must result from either a direct action of the chemical within the bladder epithelium or an indirect action not necessarily related to the tissue concentration. Recent studies have investigated the possibility of distribution in the tissues related to dose, sex and generation that might explain the bioassay data. The concentration of saccharin in the plasma and tissues of rats fed diets containing more than 5 % saccharin was higher than expected from data at lower dietary levels.[67] This probably originates from decreased excretion (see Section 3.4) rather than increased absorption, as absorption tends to be reduced at high doses (see Section 3.1). The concentration of saccharin in the bladder wall was variable but did not show evidence of excessive accumulation at high dietary levels. The concentration of saccharin in the plasma and bladder wall of adult $F_0$ and $F_1$ males was similar to or lower than the corresponding values for females.[67] Despite the slower clearance of saccharin from the foetal bladder,[62] the exposure in utero of the male bladder and other tissues was only slightly higher than that of females[73] and appeared insufficient to explain the sex specificity in tumorigenic response. Similarly, the differences in tissue concentrations of saccharin in utero in $F_1$ and $F_0$ males[67,73] appear inadequate to explain the generation-specificity of the response. Thus these tissue-distribution studies suggest that either saccharin acts by an indirect mechanism or that the immature male rat bladder shows a unique susceptibility to the presence of saccharin.

Studies on the tissue distribution of saccharin in foetal primates[74] showed a similar pattern to that reported in the foetal rat,[62,73] with only a very small amount of the total dose detected in the foetuses.

It is likely that the distribution of saccharin to human tissues may be predictable from rat data since the apparent volume of distribution after intravenous dosing in man (281 ml kg$^{-1}$)[66] is similar to that in the rat (392 ml kg$^{-1}$).[67]

### 3.3. Metabolism

Saccharin clearly differs from classical chemical carcinogens which initiate tumour development via reaction with DNA in that the parent compound is chemically stable at physiological pH and temperature and undergoes little or no metabolism either *in vivo* or *in vitro*. In high-dose two-generation cancer bioassays, however, the formation of even trace amounts of metabolites may be important in providing a possible mechanism of carcinogenicity. The metabolism of saccharin is, therefore, a key issue and warrants detailed analysis.

### 3.3.1. Animal Studies in which Metabolites were Found

Three papers have reported the presence of small amounts of radiolabelled metabolites (Fig. 7) in the excreta of animals dosed with radioactive saccharin.

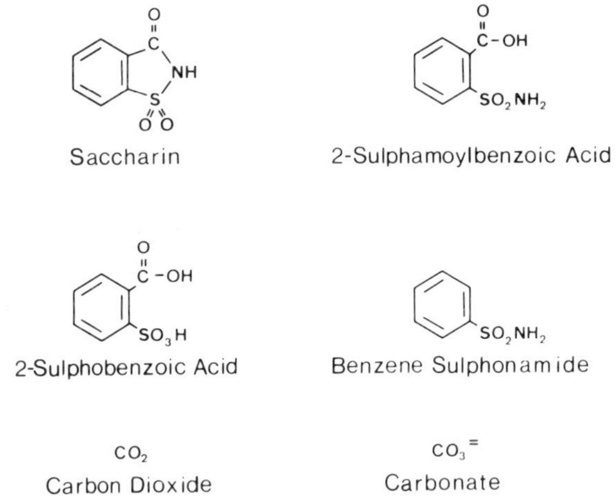

Saccharin                         2-Sulphamoylbenzoic Acid

2-Sulphobenzoic Acid              Benzene Sulphonamide

$CO_2$                            $CO_3^=$

Carbon Dioxide                    Carbonate

FIG. 7.   Saccharin and its reported 'metabolites'.

*3.3.1.1.* Kennedy *et al.*[57] gave [$^{14}$C]saccharin orally to two male and two female Charles Rivers rats at doses of 35 and 75 mg kg$^{-1}$ and 40 and 90 $\mu$Ci kg$^{-1}$. The urine was analysed by thin-layer chromatography in a solvent which gave considerable streaking of saccharin. The urine was not chromatographed directly, but was subjected to initial acidification and organic-solvent extraction (chloroform and ethyl acetate) with the possible risk of artefact formation.

The authors reported that 0·4–0·6 % and <0·1–0·6 % of the dose was excreted within 24 h as 2-sulphamoylbenzoic acid and 2-sulphobenzoate, and the totals in one rat over 72 h were 1·3 % and 0·8 %, respectively. The purity of the radiolabelled saccharin was not reported and the identity of the 'metabolites' was based on the single chromatography system. The authors also detected $^{14}CO_2$ in the expired air at low levels (0·06–0·23 %) in all animals studied, although how this could have originated was not discussed.

*3.3.1.2.* Pitkin *et al.*[64] gave [$^{14}$C]saccharin (0·04–10 mg kg$^{-1}$; 1 $\mu$Ci kg$^{-1}$) to eight female Rhesus monkeys. Analysis of the 24–48 h and 48–72 h urine by the method of Kennedy *et al.*[57] showed that a significant fraction of the radioactivity in each sample was present as 2-sulphobenzoic acid and 2-sulphamoylbenzoic acid. However, due to the rapid excretion of saccharin, the total amounts of 2-sulphobenzoic acid and 2-sulphamoyl-benzoic acid were <0·5 % and only 1–2 % of the dose, respectively. The 0–24 h urine was not analysed since Kennedy *et al.* had stated in a preliminary report that metabolites were not detected in the 0–24 h urine of saccharin-dosed rats.

In this study the dose of radioactivity used was low (1 $\mu$Ci kg$^{-1}$), which would have necessitated the chromatography of relatively large amounts of urine extracts with the subsequent possibility of artefact formation. In addition, the purity of the radiolabelled dose was not determined and the separation of the metabolites was based on a single thin-layer chromatographic solvent system.

*3.3.1.3.* Lethco and Wallace[61] produced a more thorough report on the presence of metabolites of saccharin. The [3-$^{14}$C]saccharin given to animals was purified by alumina column chromatography, although the methods used to assess the purity of saccharin were not described. 2-Sulphamoylbenzoic, but not 2-sulphobenzoic, acid was detected and the authors found both $^{14}CO_2$ in the expired air and $^{14}CO_3^{2-}$ in the urine. The levels of each of the three metabolic products were about 0·3–1 % of the dose.

The urinary products (saccharin, 2-sulphamoylbenzoic acid and carbonate) were separated by DEAE cellulose column chromatography, and their identities were confirmed by paper and thin-layer chromatography and by reverse isotope dilution. Addition of [$^{14}$C]saccharin to control urine did not result in the detection of [$^{14}$C]2-sulphamoylbenzoic acid as an artefact.

The results reported are difficult to understand for a number of reasons.

The levels of metabolites detected were similar in several mammalian species at a wide range of doses, which are not normal characteristics of enzyme-mediated reactions. Decarboxylation yielded approximately equal amounts of $CO_2$ and urinary $CO_3^{2-}$ in all species. This is unusual, since normally 95% of circulating carbonate is removed via the lungs, and suggests that the animals may have been producing alkaline urine. A possible cause for this could be the use of excess sodium hydroxide to dissolve the saccharin dose. Similarly, the use of alkali to dissolve the saccharin might have resulted in a small amount of hydrolysis to 2-sulphamoylbenzoic acid, which would explain the constant level of excretion of this 'metabolite', while addition of the pure [$^{14}C$]saccharin to control urine did not produce any artefacts.

The authors explained the presence of $^{14}CO_2$ and $^{14}CO_3^{2-}$ by proposing decarboxylation to benzenesulphonamide, which would not have been labelled and, therefore would not be detected. However, this product (benzenesulphonamide) would have been labelled in studies using [benzene ring $^{14}C$]-,[59] [$^{35}S$]-[58] or [$^3H$]-[63] saccharin, and should have been readily separable from saccharin.

Lethco and Wallace[61] concluded from their data that the similarity of metabolic profiles in the five animal species at three dose levels suggested 'that the slight breakdown of saccharin is due to simple chemical decomposition rather than an enzymic mechanism'. However, such non-enzymic processes should have been reproduced in all studies on the fate of saccharin in the body. As this was not so (see below) there may have been something 'unique' about the radiolabelled saccharin used by these workers, such as an unrecognised impurity, that was metabolised to 2-sulphamoylbenzoic acid. Byard and Golberg[59] had detected the presence of such a radioactive impurity in their [benzene ring $^{14}C$]saccharin. Renwick and Williams[75] showed that benz[d]isothiazoline-1,1-dioxide (BIT; Fig. 8), an impurity of saccharin, was metabolised via 3-hydroxy-BIT (which chromatographed similarly to saccharin) to 2-sulphamoylbenzoic acid. Thus the presence of either Byard and Golberg's impurity, BIT, or more likely 3-hydroxy-BIT in the [3-$^{14}C$]saccharin used by Lethco and Wallace[61] could result in 2-sulphamoylbenzoic acid in the urine after giving 'purified' saccharin to the animals. No explanation for the production of $^{14}CO_2$ and $^{14}CO_3^{2-}$ is apparent, but again lack of replication of the finding suggests the involvement of some unrecognised impurity.

### 3.3.2. Animal Studies that Failed to Detect Metabolism
A number of studies have failed to detect any metabolism of saccharin

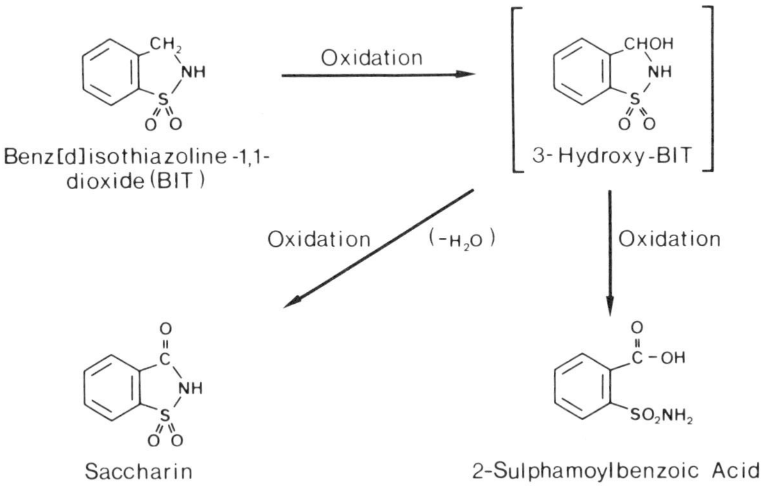

FIG. 8.   The metabolism of BIT to saccharin and 2-sulphamoylbenzoic acid (from Reference 75).

despite the use of equally sophisticated techniques and attempts to increase the extent of metabolism by induction. These studies, the methods used and the limit of detection quoted are outlined below.

*3.3.2.1.* Minegishi *et al.*[58] gave [$^{35}$S]saccharin (300 mg kg$^{-1}$; 2·3–5·5 $\mu$Ci) orally to male Wistar rats. The urine was analysed for inorganic sulphate and for saccharin metabolites by ether extraction and thin-layer chromatography (using eight different solvents). No limits of detection were given.

*3.3.2.2.* Byard[76] in an abstract, reported no metabolism of [$^{14}$C]saccharin at a dose of 40 mg kg$^{-1}$ in mice, hamsters, guinea-pigs, dogs, rats or monkeys. A full account of the studies in the rat and the monkey was published later.[59]

*3.3.2.3.* Byard and Golberg[59] administered [benzene ring $^{14}$C]-saccharin (40 mg kg$^{-1}$; 30 $\mu$Ci kg$^{-1}$) by stomach tube to four male and two female Sprague-Dawley rats and two Rhesus monkeys of each sex, and to Sprague-Dawley female rats that had been pre-treated with phenobarbitone to induce the microsomal drug-metabolising enzymes. The authors reported the presence of an impurity in the [$^{14}$C]saccharin which chromatographed similarly to saccharin, but which was excreted as a compound with similar chromatographic properties to 2-sulphamoyl-benzoic acid. If the labelled saccharin was purified prior to use, no metabolites were detected in the urine of rats or monkeys using direct chromatography. Artefacts were detected if the urine was extracted prior to chromatography, using a method similar to that of Kennedy et al.[57] but these did not correspond to the hydrolysis products, i.e. 2-sulphamoyl- and 2-sulpho-benzoic acids. The limit of detection for any chromatographic peak other than that of saccharin (i.e. metabolites) was given as 0·2%.

3.3.2.4. Matthews et al.[60] gave [3-$^{14}$C]saccharin (1 mg kg$^{-1}$ day$^{-1}$) for seven days to male Charles River rats. No metabolites were detected by thin-layer chromatography of the urine, but the limit of detection was not reported. The dose was purified by two-dimensional thin-layer chromatography. Binding of saccharin to urinary components to yield artefacts in neutral thin-layer-chromatography solvents was reported.

3.3.2.5. Ball et al.[62] gave purified [3-$^{14}$C]saccharin (16–22 mg kg$^{-1}$; 5–9 μCi rat) to female Wistar rats given normal food or diets containing 1 or 5% saccharin for up to 12 months. The excreta were analysed by chromatography and reverse isotope dilution, and specifically for $^{14}$CO$_2$ in the expired air and $^{14}$CO$_3^{2-}$ in the urine. No metabolite peaks corresponding to more than 0·3% of the dose were detected, while specific analyses for 2-sulphamoylbenzoic acid, $^{14}$CO$_2$ and $^{14}$CO$_3^{2-}$ were negative, with levels of less than 0·1, 0·03 and 0·03% of the dose, respectively. Similar results were obtained in pregnant female rats and in rabbits.

3.3.2.6. Sweatman and Renwick[63] gave purified [5-$^3$H]saccharin to male Charles Rivers rats in very low oral doses (60 μg kg$^{-1}$) and also incorporated it into a 5% saccharin diet fed to F$_1$ male rats bred, reared and maintained on a 5% saccharin diet. They also gave a low dose (400 μg kg$^{-1}$) to male rats that had been treated with 3-methylcholanthrene to increase hepatic microsomal drug metabolism. The urine was analysed for metabolites by thin-layer and high-pressure-liquid chromatography, and reverse isotope dilution for 2-sulphamoylbenzoic acid. No metabolites (>0·2% of dose) were detected.

3.3.2.7. Sweatman and Renwick[67] analysed the radioactivity in the urine of rats collected between 48 and 96 h after an intravenous dose of [$^3$H]saccharin and found that it was all unchanged saccharin.

In conclusion, studies using highly purified radiolabelled saccharin have failed to detect any metabolism despite the use of techniques of equal or higher sensitivity and specificity than those used in the positive studies outlined in Section 3.3.1. Attempts to optimise conditions for metabolism by treatment with phenobarbitone,[59] 3-methylcholanthrene,[63] chronic administration with saccharin for one[62] or two generations[63] or the use of tracer doses[63] or pregnant rats[62] have all failed to reveal any biotransformation of saccharin. In addition, saccharin is not a substrate for the microbial enzyme induced by cyclamate administration.[62] It thus appears that saccharin lacks an important characteristic of chemical carcinogens, i.e. it is not metabolised. This, combined with its lack of covalent binding to the DNA of rat liver or bladder[77] and its lack of mutagenicity has led to its classification as an epigenetic carcinogen.[78] However, it should be realised that saccharin is not unique in its metabolic stability, but is one of a group of unmetabolised, simple, acidic, cyclic sulphonamides.[79]

### 3.3.3. Human Studies

Studies on the excretion and metabolism of saccharin in humans using derivatisation and gas–liquid chromatography for quantitation, indicated that about 95 % was excreted unchanged. However, the results suggested that some saccharin was either hydrolysed to 2-sulphamoylbenzoic acid (which was then conjugated) or was bound loosely to urine constituents, since more was recovered as 2-sulphamoylbenzoic acid after alkaline hydrolysis than was measurable as saccharin itself.[80] This was confirmed using [14C]saccharin,[65] although no metabolites, conjugates or bound forms were separable by chromatography. It was reported subsequently[62] that extra saccharin could be recovered as the parent compound rather than 2-sulphamoylbenzoic acid if the urine was heated at pH 8 and then analysed by reverse isotope dilution. These data show that saccharin is not metabolised in humans, but may form a complex with an unidentified urine constituent.

### 3.4. Excretion

Since saccharin is an established food additive, most studies have concentrated on its elimination after oral administration and have shown that while most of the dose is recovered within the first 24 h, significant amounts (about 5–10 %) are excreted after this time. However, to study excretion processes for an incompletely absorbed compound like saccharin it is essential also to administer the dose by parenteral injection. Under such circumstances saccharin is eliminated more rapidly and with a greater proportion in the urine,[66,67] showing the importance of renal excretion for this compound.

### 3.4.1. Renal Elimination

Saccharin is taken up by slices of rat kidney *in vitro*, and this is an active process that works against a concentration gradient. This process can be saturated and is inhibited by *p*-aminohippurate (PAH).[70,81,82] The importance of the active secretory process *in vivo* has been demonstrated by the fact that the renal clearance of saccharin at low plasma levels is about 1·6 times that of inulin.[69,70,81] The proportion of the renal clearance occurring via renal tubular secretion may be calculated as

$$\frac{1\cdot6 - (1\cdot0 \times \alpha)}{1\cdot6}$$

where $\alpha$ is the fraction in plasma that is not protein-bound and available for filtration. Assuming there is no reabsorption from the urine (see Section 3.4.3), this gives values of 0·8 for 70 % protein-binding[67] or 0·6 for 30 % protein-binding.[69] It is therefore clear that, *in vivo*, most (60–80 %) renal elimination occurs via active, saturable renal tubular secretion. This was confirmed by the finding of 50–70 % reduction in total plasma clearance either by simultaneous administration of probenecid, an inhibitor of renal tubular secretion, or by giving very high doses of saccharin (Fig. 9).[67] A reduction in plasma clearance was found during constant intravenous infusion[67] and the plasma levels associated with reduced elimination were similar to those in rats fed 7·5 and 10 % saccharin diets *ad libitum* which developed excessive plasma and tissue accumulation.[67] This finding clearly indicates the importance of renal tubular secretion in the elimination of saccharin and the saturation of this active process that occurs at the high dietary levels used during cancer bioassays. Studies on the pharmacokinetics of intravenous saccharin in man[66] showed a similar important role for renal tubular secretion, with about 40 % decrease in clearance caused by probenecid.

### 3.4.2. Non-Renal Elimination

After an intravenous dose of labelled saccharin to rats most of the radioactivity is recovered in the urine, with only about 2 % in the faeces.[67] Since the total amount excreted in bile is only 0·3 %,[59,62] this suggests that saccharin can pass across the gut wall from plasma into the gut lumen, possibly via an active process such as has been reported for 4-nitrobenzoic acid.[83] The saccharin that enters the gut probably undergoes reabsorption

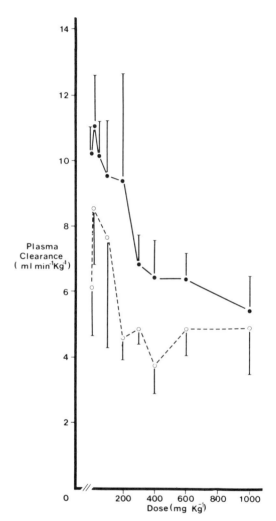

FIG. 9.   The plasma clearance of saccharin in male and female rats given various doses by intravenous administration (adapted from Reference 67).  ●, Male; ○, female.

since about 5 % of an intravenous dose enters the gut[67] but only 2 % is
recovered in the faeces.[67] As absorption from the gut is slow (see Section
3.1), this reabsorption produces a broad, very low plateau in the plasma
concentration–time curve between 24 and 72 h after an intravenous dose in
the rat[67] long after most of the dose has been recovered in the urine. No
such plateau was detected after an intravenous dose in man.[66] This is
probably due to a species difference in the extent of diffusion into the gut,
since a slow late phase is seen clearly in man after oral administration,[66,71,84] when larger amounts would be available for slow absorption
from the terminal gut (Fig. 6).

### 3.4.3. Reabsorption from the Bladder

Since renal excretion is the major route of saccharin elimination, any
reabsorption from the urinary tract could alter profoundly the efficiency of
this process, and might also determine the exposure of the bladder to
saccharin during chronic intake. A brief report that saccharin is reabsorbed
from the rat bladder extensively enough to interfere with renal clearance
measurements was therefore of considerable potential importance.[85] As
reabsorption of saccharin is negligible after microinjection into the lumen
of the rat kidney tubule,[70] extensive reabsorption would not be expected
from the bladder due to its unfavourable surface-area-to-volume ratio.
Indeed, the bladder acts as a permeability barrier so that the renal excretion
of water-soluble ionised molecules is maximised.[86] Although lipid-soluble
metabolites may be absorbed if introduced directly into the bladder,[87] they
are not normally found in bladder urine because of reabsorption in the
kidney tubule. If the normal excretory products, such as glucuronides, are
instilled into the bladder lumen reabsorption is negligible.[87]

In the report[85] which suggested that saccharin was reabsorbed from the
bladder, the ureters were left intact, and thus some 'reabsorbed' saccharin
may have been excreted back into the bladder so that the actual extent of
reabsorption (about 40 % in 2 h) may have been underestimated. This was
confirmed subsequently[88] but only if the dose was introduced via a cannula
inserted directly through the bladder wall. Since damage to the bladder
dramatically alters its permeability characteristics,[86] the study was
repeated with the dose introduced into the bladder via the ureter.[88] Under
these more physiological conditions the reabsorption of saccharin was
negligible and was not affected by feeding a 5 % saccharin diet for three
months. Thus it appears that saccharin diffuses only very slowly across the
normal bladder wall.

The concentration in the rat bladder wall after intravenous dosing was only slightly higher than that in the plasma and decreased rapidly in line with the concentrations in plasma and urine.[67] This suggests that the small amounts that enter the bladder wall, probably via the urine, are readily removed in the blood and that the bladder wall content equilibrates rapidly with changes in plasma concentration. These findings appear incompatible with suggestions[71] that the bladder wall is part of a deep, slowly equilibrating compartment, elimination from which gives rise to the slow late phase seen in the plasma concentration–time curve after oral administration. Based on the principle of pH partitioning, it has even been suggested[89] that the concentration of saccharin in the bladder wall will be 10 times that in urine. This theoretical proposal ignores the possibilities of permeability barriers and removal in plasma, and takes no account of published evidence that the value is about 100 times less![67]

## 4.    ASPARTAME

Aspartame, like saccharin and cyclamate, was discovered by serendipity during work on the synthesis of C-terminal tetrapeptide of gastrin.[90] Aspartame (aspartylphenylalanine methyl ester) is about 160 times sweeter than sucrose,[91] and a number of structural analogues show similar potency.[90]

### 4.1. Animal Studies

The fate of aspartame is less controversial than that of either cyclamate or saccharin. The molecule consists of two normal dietary amino acids linked via a peptide bond, with the carboxyl terminal group of the phenylalanine methylated (Fig. 1). An elegant study using radiolabelled aspartame has defined clearly its fate in the monkey[92] (Fig. 10). Aspartame was radiolabelled with $^{14}C$ in the methyl, aspartyl or phenylalanine moieties and $^{14}C$ in blood, urine, faeces and expired air was measured and compared with that found after administration of equimolar amounts of $[^{14}C]$methanol, $[^{14}C]$aspartic acid or $[^{14}C]$phenylalanine.

The excretion of $^{14}CO_2$ after an oral dose of $[Me\text{-}^{14}C]$aspartame was quantitatively similar to that after $[^{14}C]$methanol, but there was a time-lag of about 30 min in appearance probably due to a difference in rates of absorption. This was reflected by lower peak plasma levels of $^{14}C$ after aspartame than after methanol. The expired air was the major route of elimination of $^{14}C$ (73% in 8 h after $[^{14}C]$methanol; 67% after $[Me\text{-}^{14}C]$aspartame), with small amounts in urine (3·2% in two days after

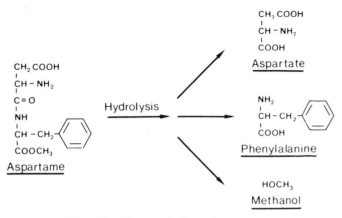

FIG. 10.   The metabolism of aspartame.

[$^{14}$C]methanol; $1\cdot6\%$ after [Me-$^{14}$C]aspartame) and negligible elimination
via the faeces. After oral doses of [Asp-$^{14}$C]aspartame and [$^{14}$C]aspartic
acid there was similar elimination of $^{14}CO_2$ in the expired air ($75\%$ and
$66\%$ of the dose in 10 h, respectively) and low excretion in urine ($2\cdot1\%$ and
$3\cdot8\%$ in four days, respectively) and in faeces ($1\cdot6\%$ and $1\cdot5\%$ in four days,
respectively). The plasma level of $^{14}$C after [Asp-$^{14}$C]aspartame was lower
than after [$^{14}$C]aspartic acid, possibly because of the delay required for
digestion within the gut lumen. Finally, comparison of the fate of [Phe-
$^{14}$C]aspartame and [$^{14}$C]phenylalanine gave almost identical results for the
rate of elimination and total amounts of $^{14}CO_2$ in 7 h ($17\%$ and $18\%$,
respectively), the elimination within two days in urine ($2\cdot8$ and $3\cdot1\%$,
respectively) and faeces ($1\cdot6$ and $4\cdot7\%$, respectively). It is clear from these
data that the fate of aspartame can be described adequately from a
knowledge of the fate of its constituent parts. The authors[92] concluded that
aspartame was probably hydrolysed in the lumen of the gut since none was
found unchanged in the plasma of aspartame-treated dogs. Hydrolysis of
dipeptides can occur also within the intestinal mucosa,[93] possibly as part of
the absorption process;[94] thus the wall of the intestine may be the main site.
In either event little or no unchanged compound reaches the portal or
systemic circulation. The total recovery of $^{14}$C after each radiolabelled
form of aspartame was less than $100\%$, the unexcreted fraction having been
incorporated into body components via normal intermediary metabolism.
It is therefore apparent that aspartame cannot be termed a true 'non-
nutritive' sweetener, but due to its intense sweetness and consequent very
low intake will provide a useful low-calorie and diabetic product.

## 4.2. Human Studies

Since aspartame is metabolised to natural endogenous dietary chemicals, its safety evaluation should be extremely simple. However, at high doses each of the three component parts of aspartame can produce toxicity in animals. Thus studies in humans have concentrated on the relationship between the elevation of blood levels of aspartate, phenylalanine and methanol after aspartame doses, in comparison with the change in blood levels after meals, and with the blood levels detected during animal toxicity studies.[95] In these studies various attempts have been made to look at 'worst case' situations, i.e. where very high levels of the hydrolysis products might be expected from either excessive dosage or an impairment to the ability of the subject to metabolise the amino acid.

*4.2.1. Likely Consumption Levels and Metabolism Studies in Man*
In the recently published final decision on aspartame[96] the Commissioner of the FDA reviewed the possible consumption of aspartame and derived the dose levels that apply to various possible situations (Table 1). It is clear

TABLE 1
THE POSSIBLE CONSUMPTION LEVELS OF ASPARTAME

| *Assumptions* | *Daily intake* $(mg\,kg^{-1}\,day^{-1})$ |
|---|---|
| Aspartame is substituted for *all sucrose* in normal diet[a] | 8 |
| Aspartame is substituted for *all carbohydrate*[b] | 25 |
| Aspartame is used in a wide variety of products including carbonated soft drinks[c]: | |
|     Age under 2 years | 6 (16) |
|     2–5 years | 11 (25) |
|     6–12 years | 6 (16) |
|     25 years and over | — (6) |
|     99th percentile for all age groups | 34 |
| Aspartame replaces 17 % of total calories (equivalent to total sucrose)[d] | 8 |
|     99th percentile | 34 |
| Aspartame replaces 50 % of total calories[e] | 22–25 |

[a,b] Based on Bureau of Foods estimates for a 60 kg adult.[96]
[c] Based on Market Research Corporation of American records of 4000 households;[96] values in parentheses are the 90th percentile.
[d,e] Based on calculations of Stegink *et al.*[95]

from the assumptions made that all values given would exceed greatly the probable intake in man. Because of its stability characteristics and bulking properties it is impossible for aspartame to replace all sucrose, let alone all carbohydrate, and thus even the lowest values given in Table 1 are overestimates. Despite this, most of the studies on changes in plasma amino-acid levels in man have used doses of 34 mg kg$^{-1}$ or more. In normal fasting individuals given a dose of 34 mg kg$^{-1}$ of aspartame there was no significant increase in the concentration of aspartate in the plasma, but the level of phenylalanine increased from 5 µmol 100 ml$^{-1}$ to about 11 µmol 100 ml$^{-1}$ [95,97] (Fig. 11). This value is well within the normal post-prandial range, and even this 99th percentile dose does not pose a risk.

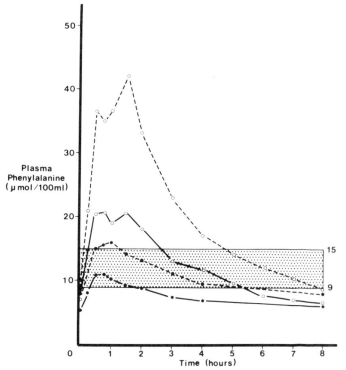

FIG. 11. The average plasma concentrations of phenylalanine after single oral doses of 34 or 100 mg kg$^{-1}$ of aspartame given to normal individuals or subjects heterozygous for phenylketonuria (PKU). The normal post-prandial range (9–15 µmol 100 ml$^{-1}$) is indicated. (Data adapted from References 97 and 101). ●—●, Normal, 34 mg kg$^{-1}$; ●--●, PKU, 34 mg kg$^{-1}$; ○—○, normal, 100 mg kg$^{-1}$; ○--○, PKU, 100 mg kg$^{-1}$.

However, not satisfied with this degree of 'over-kill' Stegink *et al.*[95] gave single oral doses of up to $200\,mg\,kg^{-1}$, which they described as 'abuse' doses. Under such extreme circumstances there was evidence of some saturation of metabolising ability since the relationship between dose and peak plasma concentrations of phenylalanine was non-linear (Fig. 12). However, the relevance of this finding (at doses equivalent to the consumption of 10–20 litres of aspartame-sweetened beverage at a single sitting) must be minimal. At these very high doses there was a small transient increase in aspartate levels in plasma, but not in erythrocytes,[98] whilst methanol was detected in the blood at low concentrations ($1$–$3\,mg$ $100\,ml^{-1}$).[99]

Neither unchanged aspartame nor the demethylated fragment, aspartylphenylalanine, were measurable in the plasma (i.e. less than $0\cdot5\,\mu mol\,100\,ml^{-1}$) even after doses as high as $50\,mg\,kg^{-1}$.[100]

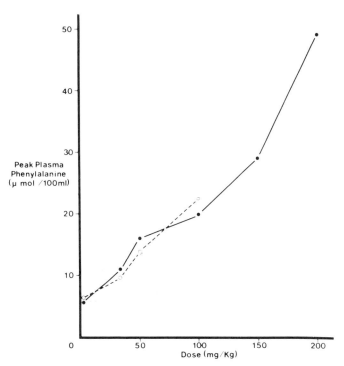

FIG. 12.   The relationship between peak plasma concentration of phenylalanine and dose of aspartame for normal adults ( ● ) and infants ( ○ ). (Data adapted from Reference 95).

### 4.2.2. Metabolism Studies under High-Risk Situations

*4.2.2.1. Infants.* It is possible that the metabolism of aspartame or its constituent amino acids may be diminished in infants to the point at which high and potentially toxic circulating blood levels may be produced. However, studies in one-year-old infants[95] have shown that the levels of circulating aspartic acid are unaffected by doses of up to 100 mg kg$^{-1}$, while the changes in phenylalanine and methanol were similar to those in adults (Fig. 12).

*4.2.2.2. Lactating women.* The possibilities of impaired metabolism of aspartame during lactation and increased amino-acid concentrations in breast milk have been suggested. However, it has been shown[100] that the handling of aspartame (50 mg kg$^{-1}$) in the mother is normal, and during the time that the concentration of phenylalanine in maternal plasma increased about four-fold (to 16 $\mu$mol 100 ml$^{-1}$) the concentration in milk increased only two- to three-fold (from 0·8 to 2·2 $\mu$mol 100 ml$^{-1}$). This suggests that even massive doses of aspartame cause variations similar to those produced by feeding.

*4.2.2.3. Heterozygotes to phenylketonuria (PKU).* Patients with phenylketonuria (PKU) are unable to metabolise phenylalanine via hydroxylation and thus have high blood levels of phenylalanine and alternative metabolic products, such as phenylpyruvic acid. This condition can lead to mental retardation if untreated, but affected individuals are usually screened soon after birth, and are given a diet low in phenylalanine. Obviously, aspartame is contraindicated for such subjects and should be avoided. The condition is inherited as a recessive trait, and the incidence of heterozygotes in the population is about 1 in 50 to 1 in 70.[101] Such individuals might be expected to have an impaired ability to metabolise phenylalanine and possibly to suffer exposure to toxic blood levels after aspartame ingestion. Studies with 34 mg kg$^{-1}$ [97] and 100 mg kg$^{-1}$ [101] have shown that such individuals do have an impaired ability to metabolise the phenylalanine component since both the peak plasma concentration and the area under the plasma concentration–time curve increased 2-fold compared with normal. However, at 34 mg kg$^{-1}$ the peak plasma level (15 $\mu$mol 100 ml$^{-1}$) only slightly exceeded the normal post-prandial rise (Fig. 11).

*4.2.2.4. In combination with monosodium glutamate.* Since the neurotoxicity of aspartate is additive to that of glutamate, which also is used as a food additive (monosodium glutamate), it is possible that a meal containing monosodium glutamate and aspartame could give toxic blood levels of these amino acids. This suggestion has been shown to be groundless, as the addition of glutamate and aspartate at $34\,mg\,kg^{-1}$ to a meal did not give blood levels of either aspartate or glutamate that were significantly higher than those caused by the test meal itself.[95]

## 5.  REGULATORY ASPECTS OF THE FATES OF NON-NUTRITIVE SWEETENERS

### 5.1.  Cyclamate

In a recent reconsideration of the use of cyclamate in the UK by the Food Additives and Contaminants Committee (FACC),[102] the Committee on Toxicity of Chemicals in Food, Consumer Products and the Environment (COT) concluded that there was no real evidence that cyclamate increased the incidence of bladder cancer in rats or lymphosarcomas in mice, and that both cyclamate and cyclohexylamine showed no clear evidence of mutagenicity. Much of the epidemiological data studied was considered to contain an estimate of the risk of both saccharin and cyclamate to man. The Committee concluded that there was no overall risk from these compounds to man, and the only positive indications involved small numbers in sub-populations, and may have represented merely chance variations within the studies. The conclusions on cyclamate metabolism were that cyclohexylamine is formed in the bowel at levels up to a maximum of 60 % of the dose, and that cyclohexylamine has a direct toxic action on the testes. However, since cyclohexylamine does not affect mice and testicular atrophy is not associated with cyclamate ingestion, its relevance to man was considered to be uncertain.[102] The COT drew attention to the co-carcinogenicity of cyclamate on the rat urinary bladder under certain experimental conditions[103] and concluded that cyclamate should be only temporarily acceptable for use in certain designated food items. However, the FACC concluded that this sweetener should not be restored to the permitted list. In view of our lack of knowledge of the mechanisms of co-carcinogenicity in test models and the fact that the only organic acids tested at very high doses have proved positive (cyclamate and saccharin), doubts must exist over the specificity of the effects seen and the significance of positive results. It would be interesting to have data in this test system on Acesulfame-K®*, an unmetabolised sweetener which was approved for use by the FACC.

* Acesulfame-K® is a registered trade mark of Hoechst AG, West Germany.

The COT did not attempt to define an ADI for cyclamate to take account of conversion to the more toxic cyclohexylamine. However, given a maximum no-effect level for cyclohexylamine in the rat and the data in Fig. 3, it is possible to derive an ADI.

Certain criteria should be considered in such an estimate:

### 5.1.1. Molecular Weight Change

The molecular weight of sodium cyclamate is 201 whilst that of cyclohexylamine is only 99. Thus each gram of cyclamate can form only 0·5 g of cyclohexylamine.

### 5.1.2. Theoretical Maximum Conversion

Cyclamate is metabolised by the gut flora at the first pass[104] and since the bacteria are associated with the human hind gut, only *unabsorbed* cyclamate is a potential substrate for conversion. Since about 40% of the dose is absorbed and excreted in the urine, only 60% is available for metabolism. This agrees well with the data given in Fig. 3.

### 5.1.3. Faecal Cyclohexylamine

Although many studies have detected cyclohexylamine in the faeces of animal and human converters of cyclamate, this is without toxicological significance, since absorbed cyclohexylamine is eliminated entirely in human urine.[48] Indeed, due to the rapid absorption of cyclohexylamine from the lower gut,[38] it is likely that much 'faecal' cyclohexylamine actually may be formed *ex vivo*.

### 5.1.4. Other Metabolites of Cyclamate and Cyclohexylamine

Cyclohexylamine is the main metabolite of cyclamate, and other metabolites such as cyclohexanol are derived from it. In both rat and man most cyclohexylamine is excreted unchanged.[48] In the rat the minor metabolites of cyclohexylamine are various aminocyclohexanols which may still possess sympathomimetic activity. In man, however, the minor metabolites are deamination products and should be relatively free from toxicity. Thus studies in the rat might overestimate cyclohexylamine toxicity in humans.

### 5.1.5. Estimation of a Realistic Conversion Value

As discussed previously, the metabolism of cyclamate to cyclohexylamine is inducible and variable. The most appropriate data to derive a usable percentage conversion value are those given in Fig. 3. Taking the average for the total population, a conversion value of about 2% would be obtained; but with the potential for up to $60\%$ metabolism in rare individuals, the usual safety factor of 100 could be reduced to only 3. Such an approach is clearly invalid and inappropriate. Conversely, if a conversion value of $60\%$ was applied, this would give a 100-fold safety margin for the 99th percentile of conversion, and this is overcautious. Such an approach was not applied to the aspartame data and judgement[96] (see Section 5.3).

The 100-fold safety factor is designed to protect the public from differences between the test species and man (10-fold) and between different individuals (10-fold). Individual differences can exist in both the metabolism of the compound and in susceptibility to its effects, and it would seem sensible to divide the latter 10-fold factor into two equal weightings of approximately 3-fold, and to use this factor to help interpret the vast amount of data relating to the metabolism of cyclamate in man. Since the *maximum* conversion of cyclamate is $60\%$ it would be logical to apply the 3-fold factor to this value to obtain a *working conversion value of $20\%$*. It is interesting to note that the WHO[105] recently arrived at a formula which incorporated a $30\%$ conversion value but this was then applied to the $60\%$ unabsorbed so that the 'effective conversion' was $18\%$. (This appears to the right answer but for the wrong reasons!)

Using the conversion value of $20\%$ derived above and the population analysis given in Fig. 3, it is possible to calculate the safety factors that apply for different percentage conversions. These are given in Table 2 but it must be emphasised that they are the safety factors that remain *after* one of the major variables of extrapolation (individual differences in metabolism) has been taken into account. Thus it appears that the use of a $20\%$ conversion value would act as an adequately cautious safeguard, with $96\%$ of the population still having a 100-fold or better safety factor. Indeed, there are some indications that this may represent an overconservative and unnecessarily restrictive value.

The metabolism of cyclamate requires continuous high doses both to induce and to maintain the metabolising system, since the ability to metabolise is lost rapidly on cessation of high-dose daily intake. Thus it is likely that the experimentally derived data used for Fig. 3 overestimate the extent of conversion during normal dietary intake. It is particularly interesting to note the extent of cyclohexylamine excretion measured in 50 Japanese subjects during a 24-h period in the summer of 1969, immediately prior to the ban.[29] The results (Table 3) show that, even during extensive use of cyclamate as a food additive, only one individual excreted cyclohexylamine at a level of more than about $0.5\,mg\,kg^{-1}\,day^{-1}$. In addition, long-term studies in man[106] and rat[26] suggest that even extensive

214 A. G. RENWICK

## TABLE 2
### SAFETY FACTORS IN RELATION TO CYCLAMATE METABOLISM

| Percentage excreted as cyclohexylamine in urine[a] | Percentage of population[a] | Safety factor[b] |
|---|---|---|
| 0–0·1 | 73 | Infinity–20 000 |
| 0·1–1 | 13 | 20 000–2 000 |
| 1–20 | 10 | 2 000–100 |
| 20–60 | 3 | 100–33 |
| 60+ | 1 | < 33 |

[a] Data obtained during chronic experimental intake (Fig. 3) referring to both males and females who are not at risk.
[b] Safety factor remaining for interspecies differences and individual variations in susceptibility to cyclohexylamine if a 20 % conversion figure is used in conjunction with a 100-fold safety factor (applied to the maximum no-effect level for cyclohexylamine in the rat). Assuming that the no-effect level for cyclohexylamine is about $100\,mg\,kg^{-1}\,day^{-1}$, then the ADI for *cyclamate* would be:

$$\frac{100\,mg\,kg^{-1}\,day^{-1}}{100} \times \frac{100}{20} \times 2 = 10\,mg\,kg^{-1}\,day^{-1} \text{ in man}$$

where, 100 = normal safety factor; 100/20 = factor to allow for only 20 % of the dose of cyclamate being converted; and 2 = molecular weight factor.

## TABLE 3
### EXCRETION OF CYCLOHEXYLAMINE DURING DIETARY INTAKE OF CYCLAMATE[a]

| Cyclohexylamine (milligrams excreted per day) | Number of subjects |
|---|---|
| 0–1 | 31 |
| 1–10 | 10 |
| 10–20 | 4 |
| 20–30 | 3 |
| 30+ | 1 (129 mg) |

[a] Data refer to 50 Japanese subjects who consumed a normal diet. The study[29] was conducted during August and September, 1969, when cyclamate was used extensively as a food additive.

converters are unlikely to retain a high metabolising capacity for prolonged periods during chronic intake of cyclamate, so that 'lifetime protection' by the use of 20% conversion and a 100-fold safety factor may be unnecessary and unrealistic.

A number of other factors are implicit in the use of a simple 20% conversion factor for cyclamate metabolism in man. It is assumed that the relationship between dose and percentage conversion is constant, yet evidence suggests that high doses may be metabolised less extensively.[9] Also, it is assumed that feeding cyclohexylamine in the diet is a good predictor of the exposure when it is formed from cyclamate. However, a knowledge of the pharmacokinetics of cyclamate and cyclohexylamine suggests that feeding cyclohexylamine may result in short periods with much higher blood levels. Thus the direct-feeding study may not be ideal and possibly will tend to overestimate cyclohexylamine toxicity.

In conclusion there is probably more information about individual variation in the fate of cyclamate in man, than about any other drug or food additive. Such information should be the basis for the estimation of an ADI via a percentage conversion figure that will adequately protect the public. A value of 20% appears prudent at this time, but this may be considerably overcautious in protecting the public from any possible effects of cyclohexylamine during lifetime exposure to dietary sources.

For the *logical* regulation of cyclamate, considerably more data of the type provided by Asahina et al.[29] would be useful. The continued use of cyclamate in a number of countries with different patterns and levels of intake provides an opportunity to measure actual cyclohexylamine excretion in normal cyclamate users. Such data could be useful directly in conjunction with cyclohexylamine toxicity data and a suitable safety factor, to allow the establishment of safe patterns of use of cyclamate without involving a contentious, and to some extent arbitrary, conversion factor. However, for countries contemplating the re-introduction of cyclamate, the use of a reasonably logical conversion value (20%) should allow a safe level of use. Monitoring of cyclohexylamine formation could then determine whether this was too liberal or too conservative (as appears more likely).

## 5.2. Saccharin

The FACC[102] also considered the available data on saccharin and concluded that administration at high dietary levels increased the incidence of bladder tumours in rats when given for two generations. Their analysis of the epidemiology data suggested that 'saccharin [was] neither a complete carcinogen nor a bladder tumour promoter in humans'. Thus they clearly pinpointed the discrepancy between animal and human data. The COT of FACC recommended only temporary acceptance for the use of saccharin in food. The report drew attention to the fact that the mechanism of tumorigenicity is not known, but it appears to affect only males and only at high doses. These conclusions are similar to those reached in the USA,[72] although in the US report the metabolism and pharmacokinetics were discussed in considerably more detail.

The apparent discrepancy between the positive animal carcinogenicity data and the overall negative human epidemiology evidence could arise from various sources:

1. *Inadequate sensitivity in the epidemiology studies.* Most studies have been extensive and have utilised large numbers of patients and matched controls. However, it is possible that the period of intensive exposure has not been long enough to allow for any possible latency period. In addition, dietary histories will tend to be biased towards recent memory. Certain groups of individuals have been shown to have an increased risk of bladder cancer due to their use of artificial sweeteners, but there is little consistency among the studies. These positive findings may be random, spurious associations,[102] or they may be at the limit of detection of the epidemiological methods.

2. *Excessive sensitivity in the animal bioassay data.* It is possible that the use of high doses in animal studies, designed to increase sensitivity of the model system, may overload the normal physiological and biochemical processes to an extent where there is a non-linear relationship between dose and response. Under such circumstances, effects seen only when the animal has overt signs of toxicity may be of limited relevance to much lower human exposure levels. For this reason the concept of a maximum tolerated dose (MTD) is used. This is usually taken as the dose which produces a 20% decrease in body weight. However, this 'rule of thumb' has little physiological or biochemical basis, and an MTD should be derived for each compound based on its fate in the body and effects on biochemical, physiological and immunological processes.

3. *A species difference.* The animal data may be of limited relevance to man if there is a clear species difference in either: (1) the processes by which the chemical produces its effects; or (2) the sensitivity of the target organ to those processes.

The extensive knowledge of the fate of saccharin in animals and man has tended to confirm suspicions that at the present time two-generation high-dose studies may be investigating the experimental protocol as much as the test substance. Key points in this regard are that saccharin is not metabolised (see Section 3.3), does not bind to DNA[77] and passes through the body unchanged. Thus it cannot act as a simple electrophilic carcinogen and the mechanism of tumorigenicity must be indirect. This suggests that the one-hit carcinogenesis philosophy, which is behind the Delaney Amendment to the Food and Cosmetics Act in the USA, is inappropriate in the case of saccharin. The National Academy of Sciences (NAS)[72] reviewed the available data on the fate of saccharin *in vivo* and considered that they were equivocal and that there may be some metabolism. This appears incorrect after careful consideration of all the data; saccharin must be regarded as an example of an unmetabolised compound.[79] (It is interesting to note that both of the reported 'metabolites' are, like saccharin, acidic compounds of low toxicity.[107])

Implications that the bladder may be the target for carcinogenesis due to extensive accumulation of saccharin, especially *in utero* during chronic administration, appeared logical at the time of the NAS review, but recent findings[67,73] have rendered this suggestion untenable. In addition, no clear sex difference in tissue exposure has emerged. All these findings suggest that saccharin acts at high doses via some indirect sex-specific mechanism.

Studies of the pharmacokinetics of saccharin in the rat and man have suggested that the rat is a reasonable model for man, showing slow absorption, a similar apparent volume of distribution and rapid elimination, primarily via active renal tubular secretion. However, it is clear[67] that the carcinogenic doses given to rats saturate the renal tubular secretion capacity for saccharin. This active mechanism has a broad substrate specificity and its saturation during gestation and lactation could alter the exposure of the offspring to normal endogenous substrates throughout this critical period. Studies using the model compound *p*-aminohippurate as a second substrate have shown reversible, competitive inhibition of uptake *in vitro*[82] by high dietary levels of saccharin, and such competition occurs for endogenous substrates *in vivo*. Recently, Sims and Renwick[108] have demonstrated saturation of elimination of certain metabolites of tryptophan (which is a promoter of bladder carcinogens) in rats fed carcinogenic dietary levels of saccharin. Thus it appears that the pharmacokinetics of saccharin at high doses may alter profoundly the biochemistry and physiology of the test animal. Such findings, combined with the overall negative human epidemiology data, suggest that criteria other than simple weight-loss may be more appropriate to assess the maximum dose to be used in animal toxicity and carcinogenicity studies. As noted by the FACC,[102] when granting temporary acceptance for saccharin, there are a number of interesting studies underway at present including a two-generation dose–response cancer bioassay as well as highly relevant studies on tumour promotion, mineral balance and other biochemical and physiological effects. It is to be hoped that with these results we shall be able to extrapolate an accurate dose response from animals to human exposure levels. In addition, we should be armed with an understanding of some of the other consequences of feeding very high dietary levels to test animals. A rational use of animal data must be the aim of all thinking toxicologists[109] and in this respect the recent decision by the Commissioner of the FDA in relation to aspartame[96] is a model of logic and common sense.

### 5.3. Aspartame

The regulation of aspartame is simpler than that of either saccharin or cyclamate mainly because it is metabolised to endogenous or naturally occurring compounds. The decisions were eased by the study of the fate of 'abuse' doses in humans, combined with the known increases in blood levels of aspartate and phenylalanine after normal meals. The potential toxicological problems associated with aspartame relate to the possibility of brain damage from phenylalanine or aspartate, brain tumours and also opthalmological damage due to the liberation of methanol.

The increase in the plasma level of phenylalanine after a dose of $34 \, mg \, kg^{-1}$ of aspartame (the 99th percentile) does not significantly exceed the normal post-prandial rise, even when given to subjects who are heterozygous for phenylketonuria. This is to be expected since the phenylalanine content of a 4-oz cooked hamburger is 1150 mg or about $20 \, mg \, kg^{-1}$.[96] It would therefore appear logical that if aspartame is an unacceptable risk so too are hamburgers! Such simplistic logic did not deter critics of aspartame who pointed out the risk to undiagnosed PKU infants and pregnant women with hyperphenylalaninaemia. However, the former group are protected because aspartame is not being approved for infant formulas, while for the latter group the major risk arises from the condition being unrecognised and the subjects consuming phenylalanine in the normal diet. The introduction of aspartame was considered not to pose a significant risk of diffuse brain damage associated with phenylalanine.[96]

A similar rationale was applied to the problems of the aspartate component. A review of the animal data[96] suggested that a blood level of glutamate plus aspartate equivalent to at least $100 \mu mol$ $100 ml^{-1}$ was necessary to produce symptoms of neurotoxicity even in the most sensitive of animal species. The normal circulating blood level is equivalent to a total of about $5 \mu mol$ $100 ml^{-1}$ under fasting conditions, and this rises to a maximum of $8.8 \mu mol$ $100 ml^{-1}$ after a meal or $11.7 \mu mol$ $100 ml^{-1}$ after the meal plus monosodium glutamate (MSG) and aspartame at $34 mg kg^{-1}$.[95] It is thus inconceivable[96] that the blood levels of these amino acids could be made to approach toxic levels by the consumption of aspartame, especially since aspartame alone causes no statistically significant net increase in aspartate levels at doses of less than $50 mg kg^{-1}$.[95] In addition aspartame at $34 mg kg^{-1}$ did not increase further the total concentration of these amino acids that could be produced by a $50 mg kg^{-1}$ dose of MSG. Indeed, from the data discussed in the Commission's Report,[96] it would appear that MSG is probably a greater cause for concern than is aspartame.

However, it is possible to analyse the published data slightly differently and to reach a different conclusion. After giving a single dose of $34 mg kg^{-1}$ of aspartame alone the total fasting aspartate plus glutamate levels increased from about $3.1$ to $4.7 \mu mol$ $100 ml^{-1}$ in normal subjects and from $4.6$ to about $6.0 \mu mol$ $100 ml^{-1}$ in PKU heterozygotes. Thus this dose (equivalent to a 99th percentile of intake) caused an increase of about $1.5 \mu mol$ $100 ml^{-1}$ in the plasma concentrations towards a 'minimum *effect level*' in animals of $100 \mu mol$ $100 ml^{-1}$. This is equivalent to a '*non*-safety factor' of about 66 at the 99th percentile of intake, and this factor must still allow for individual variations in both metabolism and sensitivity. This analysis was not intended to demonstrate that the FDA conclusion was incorrect and that aspartame poses a risk. Rather it draws attention to the difference in regulatory philosophy that seems to apply to 'endogeneous toxins' like phenylalanine, aspartate and glutamate in this context and 'exogenous toxins' like cyclamate and its metabolite cyclohexylamine at the 99th percentile of metabolism. Similarly, concern has been expressed about the promotion of bladder carcinogens by very high dietary levels of cyclamate and saccharin ($5\%$ or more). However, the strongly positive result with the essential amino acid tryptophan at only 10 times the normal dietary level has received comparatively little attention or interest.

The problem of a possible relationship between aspartame ingestion and brain tumours was considered by the FDA[96] and is a matter of simple (?) toxicology, unconfounded by metabolism or blood levels.

Finally, the potential of the methanol formed from aspartame to cause damage, especially to the eye, has been studied in man at an 'abuse' dose and it has been concluded that the amounts produced could not be of any toxicological significance.[95,99]

### 5.4. Other Sweeteners

Acesulfame-K is a simple heterocyclic compound that is between 130[102] and 200[110] times sweeter than sucrose. It has recently received approval by the FACC[102] and is under review by most regulatory authorities. Few data on this sweetener have been published[110] and the FACC decision was based on unpublished reports provided by the manufacturers. The published information[102,110] suggests that Acesulfame-K is excreted mostly in the urine without undergoing detectable metabolism. However, from the FACC report[102] it would appear that this compound[110] shares a number of characteristics with saccharin, such as an acid, heterocyclic structure, excretion without metabolism by the tissues and also the tendency to cause caecal enlargement. In view of these similarities it would be extremely interesting to know the answers to certain questions: e.g. is its metabolism induced by chronic administration? Does it promote the bladder carcinogenicity of known carcinogens such as $N$-methyl-$N$-nitrosourea (MNU) and $N$-[4-(5-nitro-2-furyl)-2-thiazolyl]formamide (FANFT)? Finally, does it produce bladder tumours in rats when fed at 5 or 7·5 % in the diet for two generations? The answers to these questions are not given clearly in the FACC report, and until this information is available it appears that the view of the Committee with respect to the sweeteners discussed in this chapter can be summarised as 'better the devil you don't know, than the devil you do'. Indeed, in view of the extensive epidemiology data pertaining to cyclamate and saccharin, the elevated status of Acesulfame-K appears questionable.

## REFERENCES

1. AUDRIETH, L. F. and SVEDA, M. (1944). *J. Org. Chem.*, **9**, 89.
2. WILLIAMS, R. T. (1959). *Detoxication Mechanisms*, 2nd edn., Chapman and Hall, London.
3. WEEDON, B. C. L. (1970). *Chem. Brit.*, **6**, 242.
4. PRICE, J. M. *et al.* (1969). *Science*, **167**, 1131.
5. OSER, B. L., CARSON, S., COX, G. E., VOGIN, E. E. and STERNBERG, S. S. (1975). *Toxicology*, **4**, 315.
6. TAYLOR, J. P., RICHARDS, R. K. and DAVIN, J. C. (1951). *Proc. Soc. Exp. Biol. Med.*, **78**, 530.
7. MILLER, J. P., CRAWFORD, L. E. M., SONDERS, R. C. and CARDINAL, E. V. (1966). *Biochem. Biophys. Res. Commun.*, **25**, 153.

8. RENWICK, A. G. and WILLIAMS, R. T. (1972). *Biochem. J.*, **129**, 869.
9. COLLINGS, A. J. (1971). In: *Sweetness and Sweeteners*, G. C. Birch, L. F. Green and C. B. Coulson (Eds.), Applied Science Publishers Ltd, London, 51.
10. PAREKH, C., GOLBERG, E. K. and GOLBERG, L. (1970). *Toxicol. Appl. Pharmacol.*, **17**, 282.
11. WALLACE, W. C., LETHCO, E. J. and BROWER, E. A. (1970). *J. Pharmacol. Exp. Ther.*, **175**, 325.
12. KOJIMA, S., ICHIBAGASE, H. and IGUCHI, S. (1966). *Chem. Pharm. Bull. (Tokyo)*, **14**, 959.
13. KOJIMA, S., ICHIBAGASE, H. and IGUCHI, S. (1966). *Chem. Pharm. Bull. (Tokyo)*, **14**, 965.
14. SCHOENBERGER, J. A., RIX, D. M., SAKAMOTO, A., TAYLOR, J. D. and KARK, R. M. (1953). *Amer. J. Med. Sci.*, **225**, 551.
15. DAVIS, T. R. A., ADLER, N. and OPSAHL, J. C. (1969). *Toxicol. Appl. Pharmacol.*, **15**, 106.
16. SONDERS, R. C. and WIEGAND, R. G. (1968). *Toxicol. Appl. Pharmacol.*, **12**, 291.
17. KOJIMA, S. and ICHIBAGASE, H. (1968). *Chem. Pharm. Bull. (Tokyo)*, **16**, 1619.
18. ICHIBAGASE, H., IMAMURA, I., KINOSHITA, A. and KOJIMA, S. (1972). *Chem. Pharm. Bull. (Tokyo)*, **20**, 947.
19. BICKEL, M. H., BURKARD, B., MEIER-STRASSER, E. and VAN DEN BROEK-BOOT, M. (1974). *Xenobiotica*, **4**, 425.
20. PITKIN, R. M., REYNOLDS, W. A. and FILER, L. J. (1969). *Proc. Soc. Exp. Biol. Med.*, **132**, 993.
21. KOJIMA, S. and ICHIBAGASE, H. (1966). *Chem. Pharm. Bull. (Tokyo)*, **14**, 971.
22. OSER, B. L., CARSON, S., VOGIN, E. E. and SONDERS, R. C. (1968). *Nature*, **220**, 178.
23. PROSKY, L. and O'DELL, R. G. (1971). *J. Pharm. Sci.*, **60**, 1341.
24. DALDERUP, L. M., KELLER, G. H. M. and SCHOUTEN, F. (1970). *Lancet*, **1**, 845.
25. TESORIERO, A. A. and ROXON, J. J. (1975). *Xenobiotica*, **5**, 25.
26. RENWICK, A. G. (1977). In: *Drug Metabolism—From Microbes to Man*, D. V. Parke and R. L. Smith (Eds.), Taylor and Francis, London, 169.
27. ASAHINA, M., YAMAHA, T., SARRAZIN, G. and WATANABE, K. (1972). *Chem. Pharm. Bull. (Tokyo)*, **20**, 102.
28. ICHIBAGASE, H., KOJIMA, S., SUENAGA, A. and INOUE, K. (1972). *Chem. Pharm. Bull (Tokyo)*, **20**, 1093.
29. ASAHINA, M., YAMAHA, T., WATANABE, K. and SARRAZIN, G. (1971). *Chem. Pharm. Bull. (Tokyo)*, **19**, 628.
30. PAWAN, G. L. S. (1970). *Proc. Nutr. Soc.*, **20**, 10A.
31. HENGSTMANN, J., EICHELBAUM, M. and DENGLER, H. J. (1971). *Arch. Pharmakol.*, **270**, Suppl. R60.
32. GLOGNER, P. (1970). *Humangenetik*, **9**, 230.
33. LEAHY, J. S., TAYLOR, T. and RUDD, C. J. (1967). *Food Cosmet. Toxicol.*, **5**, 595.
34. BLUMBERG, A. G. and HEATON, A. M. (1970). *J. Chromatog.*, **48**, 565.
35. LITCHFIELD, M. H. and SWAN, A. A. B. (1971). *Toxicol. Appl. Pharmacol.*, **18**, 535.

36. LEAHY, J. S., WAKEFIELD, M. and TAYLOR, T. (1967). *Food Cosmet. Toxicol.*, **5**, 447.
37. WILLS, J. H., JAMESON, E., STOEWSAND, G. and COULSTON, F. (1968). *Toxicol. Appl. Pharmacol.*, **12**, 292.
38. DRASAR, B. S., RENWICK, A. G. and WILLIAMS, R. T. (1972). *Biochem. J.*, **129**, 881.
39. ASAHINA, M., NIIMURA, T., YAMAHA, T. and TAKAHASHI, T. (1972). *Agric. Biol. Chem.*, **36**, 711.
40. GOLBERG, L., PAREKH, C., PATTI, A. and SOIKE, K. (1969). *Toxicol. Appl. Pharmacol.*, **14**, 654.
41. SONDERS, R. C., NETWAL, J. C. and WIEGAND, R. G. (1969). *Pharmacologist*, **11**, 241.
42. KOJIMA, S. and ICHIBAGASE, H. (1968). *Chem. Pharm. Bull. (Tokyo)*, **16**, 1851.
43. SUENAGA, A., KOJIMA, S. and ICHIBAGASE, H. (1972). *Chem. Pharm. Bull. (Tokyo)*, **20**, 1357.
44. BERNHARD, K. (1937). *Hoppe-Seylers Z. Physiol. Chem.*, **248**, 256.
45. Editorial Comment (1967). *Food Cosmet. Toxicol.*, **5**, 398.
46. ELLIOTT, T. H., LEE-YOONG, N. Y. and TAO, R. C. C. (1968). *Biochem. J.*, **109**, 11P.
47. SONDERS, R. C., ESTEP, C. B. and WIEGAND, R. G. (1968). *Fed. Proc.*, **27**, 238.
48. RENWICK, A. G. and WILLIAMS, R. T. (1972). *Biochem. J.*, **129**, 857.
49. KUREBAYASHI, H., TANAKA, A. and YAMAHA, T. (1979). *Biochem. Pharmacol.*, **28**, 1719.
50. NIIMURA, T., TOKEIDA, T. and YAMAHA, T. (1974). *J. Biol. Chem.*, **75**, 407.
51. BENSON, G. A. and SPILLANE, W. J. (1976). *J. Pharm. Sci.*, **65**, 1841.
52. SPILLANE, W. J. and BENSON, G. A. (1978). *J. Pharm. Sci.*, **67**, 226.
53. SPILLANE, W. J., BENSON, G. A. and McGLINCHEY, G. (1979). *J. Pharm. Sci.*, **68**, 372.
54. KLAVERKAMP, J. F. and DIXON, R. L. (1969). *Proc. Soc. Western Pharmacol. Soc.*, **12**, 75.
55. HETTLER, H. (1973). *Adv. Heterocyclic Chem.*, **15**, 233.
56. DE GARMO, O., ASHWORTH, G. W., EAKER, C. M. and MUNCH, R. H. (1952). *J. Amer. Pharm. Assoc.*, **41**, 17.
57. KENNEDY, G., FANCHER, O. E., CALANDRA, J. C. and KELLER, R. E. (1972). *Food Cosmet. Toxicol*, **10**, 143.
58. MINEGISHI, K., ASAHINA, M. and YAMAHA, T. (1972). *Chem. Pharm. Bull. (Tokyo)*, **20**, 1351.
59. BYARD, J. L. and GOLBERG, L. (1973). *Food Cosmet. Toxicol.*, **11**, 391.
60. MATTHEWS, H. H., FIELDS, M. and FISHBEIN, L. (1973). *J. Agric. Food Chem.*, **21**, 916.
61. LETHCO, E. J. and WALLACE, W. C. (1975). *Toxicol.*, **3**, 287.
62. BALL, L. M., RENWICK, A. G. and WILLIAMS, R. T. (1977). *Xenobiotica*, 7, 189.
63. SWEATMAN, T. W. and RENWICK, A. G. (1979). *Science*, **205**, 1019.
64. PITKIN, R. M., ANDERSEN, D. W., REYNOLDS, W. A. and FILER, L. J. (1971). *Proc. Soc. Exp. Biol. Med.*, **137**, 803.
65. BYARD, J. L., McCHESNEY, E. W., GOLBERG, L. and COULSTON, F. (1974). *Food Cosmet. Toxicol.*, **12**, 175.

66. SWEATMAN, T. W., RENWICK, A. G. and BURGESS, C. D. (1981). *Xenobiotica*, **11**, 531.
67. SWEATMAN, T. W. and RENWICK, A. G. (1980). *Toxicol. Appl. Pharmacol.*, **55**, 18.
68. AGREN, A. and BACK, T. (1973). *Acta Pharm. Suecica*, **10**, 223.
69. BEKERSKY, I., POYNOR, W. J. and COLBURN, W. A. (1980). *Drug Metab. Disp.*, **8**, 64.
70. BOURGOIGNIE, J. J., HWANG, K. H., PENNELL, J. P. and BRICKER, N. S. (1980). *Amer. J. Physiol.*, **238**, F10.
71. COLBURN, W. A., BEKERSKY, I. and BLUMENTHAL, H. P. (1981). *J. Clin. Pharmacol.*, **21**, 147.
72. NATIONAL ACADEMY OF SCIENCE (1978). *Saccharin: Technical Assessment of Risks and Benefits*, Report No. 1, Washington, DC.
73. SWEATMAN, T. W. and RENWICK, A. G. (1982). *Toxicol. Appl. Pharmacol.*, **62**, 465.
74. PITKIN, R. M., REYNOLDS, W. A., FILER, L. J. and KLING, T. G. (1971). *Amer. J. Ob. Gynecol.*, **111**, 280.
75. RENWICK, A. G. and WILLIAMS, R. T. (1978). *Xenobiotica*, **8**, 475.
76. BYARD, J. L. (1972). *Toxicol. Appl. Pharmacol.*, **22**, 291.
77. LUTZ, W. K. and SCHLATTER, C. (1977). *Chem.-Biol. Interactions*, **19**, 253.
78. ASHBY, J., STYLES, J. A., ANDERSON, D. and PATON, D. (1978). *Food Cosmet. Toxicol.*, **16**, 95.
79. RENWICK, A. G. (1983). In: *Biological Basis of Detoxication*, J. Caldwell and W. B. Jakoby (Eds.), Academic Press, New York, 151.
80. MCCHESNEY, E. W. and GOLBERG, L. (1973). *Food Cosmet. Toxicol.*, **11**, 403.
81. GOLDSTEIN, R. S., HOOK, J. B. and BOND, J. T. (1978). *J. Pharmacol. Exp. Ther.*, **204**, 690.
82. BERNDT, W. O., REDDY, R. V. and HAYES, A. W. (1981). *Toxicology*, **21**, 305.
83. GARDNER, D. M. and RENWICK, A. G. (1978). *Xenobiotica*, **8**, 679.
84. COLBURN, W. A., BEKERSKY, I. and BLUMENTHAL, H. P. (1981). *Clin. Pharmacol. Therap.*, **30**, 558.
85. COLBURN, W. A. (1978). *J. Pharm. Sci.*, **67**, 1493.
86. LAWSON, T. A. (1978). In: *Health and Sugar Substitutes*, B. Guggenheim (Ed.), S. Karger, Basel, 48.
87. OGLESBY, L. A., FLAMMANG, T. J., TULLIS, D. L. and KADLUBAR, F. F. (1981). *Carcinogenesis*, **2**, 15.
88. RENWICK, A. G. and SWEATMAN, T. W. (1979). *J. Pharm. Pharmacol.*, **31**, 650.
89. WADDELL, W. J. and LACHANCE, M. P. (1981). *Science*, **214**, 143.
90. MAZUR, R. H., SCHLATTER, J. M. and GOLDKAMP, A. H. (1969). *J. Amer. Chem. Soc.*, **91**, 2684.
91. CLONINGER, M. R. and BALDWIN, R. E. (1970). *Science*, **170**, 81.
92. OPPERMANN, J. A., MULDOON, E. and RANNEY, R. E. (1973). *J. Nutr.*, **103**, 1454.
93. HEIZER, W. D. and LASTER, L. (1969). *Biochim. Biophys. Acta*, **185**, 409.
94. ADDISON, J. M. *et al.* (1975). *Clin. Sci. Mol. Med.*, **49**, 313.
95. STEGINK, L. D., FILER, L. J., BAKER, G. L., BRUMMEL, M. C. and TEPHLY, T. R. (1979). In: *Health and Sugar Substitutes*, B. Guggenheim (Ed.), S. Karger, Basel, 160.

96. *Aspartame: Commissioners Final Decision* (1981). Department of Health and Human Services, Public Health Service, Food and Drug Administration, (Docket No. 75F-0355), Washington, DC.

97. STEGINK, L. D., FILER, L. J., BAKER, G. L. and MCDONNELL, J. F. (1979). *J. Nutr.*, **109**, 708.

98. STEGINK, L. D., FILER, L. J. and BAKER, G. L. (1981). *J. Toxicol. Environ. Health*, **7**, 291.

99. STEGINK, L. D., FILER, L. J. and BAKER, G. L. (1981). *J. Toxicol. Environ. Health*, **7**, 281.

100. STEGINK, L. D., FILER, L. J. and BAKER, G. L. (1979). *J. Nutr.*, **109**, 2173.

101. STEGINK, L. D., FILER, L. J., BAKER, G. L. and MCDONNELL, J. E. (1980). *J. Nutr.*, **110**, 2216.

102. *Food Additives and Contaminants Committee Report on the Review of Sweeteners in Food* (1982). (FAC/REP/34), HMSO, London.

103. HICKS, R. M. (1980). *Brit. Med. Bull.*, **36**, 39.

104. RENWICK, A. G. (1982). In: *Presystemic Drug Elimination*, C. F. George, D. G. Shand and A. G. Renwick (Eds.), Butterworths, London, 3–28.

105. WORLD HEALTH ORGANISATION. (1978). *Twenty-First Report of the Joint FAO/WHO Expert Committee on Food Additives*, Technical Report Series 617.

106. WILLS, J. H., SERRONE, D. M. and COULSTON, F. (1981). *Regulatory Toxicol. Pharmacol.*, **1**, 163.

107. KENNEDY, G. L., FANCHER, O. E. and CALANDRA, J. C. (1976). *Toxicol.*, **6**, 133.

108. SIMS, J. and RENWICK, A. G. (1983). *Toxicol. Appl. Pharmacol.*, (In press).

109. SQUIRE, R. A. (1981). *Science*, **214**, 877.

110. ARPE, H. J. (1979). In: *Health and Sugar Substitutes*, B. Guggenheim (Ed.), S. Karger, Basel, 178.

*Chapter 8*

# NON-NUTRITIVE SWEETENERS IN FOOD SYSTEMS

M. G. Lindley

*Tate & Lyle Group Research and Development, Reading, UK*

## SUMMARY

*Non-nutritive sweeteners have been available for use in foods for many years, but the recent concern about obesity and dental caries, as well as an increasing appreciation of the need for special diets, principally in the management of diabetes, has resulted in the creation of many new food products to satisfy these needs. To incorporate non-nutritive sweeteners in foods and beverages successfully, a thorough understanding of their roles in each food application is necessary. The sensory and chemical properties of the non-nutritive sweeteners must be understood as these frequently will restrict application. Finally, attention must be paid to many other factors, including texture, bulk, flavour quality and carbonation levels.*

*This chapter reviews the properties of nutritive sweeteners in foods and discusses the sensory and chemical properties of permitted non-nutritive sweeteners. Factors which will govern the acceptability of non-nutritive-sweetened foods are assessed, and examples of the successful incorporation of non-nutritive sweeteners into foods are given. Finally, the impact of new developments on the use of non-nutritive sweeteners in food systems is discussed.*

## 1. INTRODUCTION

The demand for non-nutritive sweeteners in food systems is almost invariably in response to the need for special diets, as in the management of

225

diabetes or sucrose-intolerance, or to consumer concern about obesity and dental caries. Whether the initiative comes from consumers directly or only via self-appointed 'consumer protectionists' is open to question. For the food manufacturer, saccharin represents a low-cost alternative to sucrose, although it is crucial to bear in mind that if the manufacturer's goal is the substantial replacement of nutritive sweeteners, it must be based on a clear understanding of the exact role(s) of the nutritive sweetener in each particular application. In particular, a clear distinction between the use of a nutritive sweetener for its sweetening power and for its bulking, textural and preservative properties must be appreciated. Non-nutritive sweeteners find a wide and relatively straightforward application in food systems where water is used as the principal bulking agent. However, there are many food categories where a need for a low- or reduced-calorie version has been recognised but which do not have water as the principal bulking agent. In order to manufacture a satisfactory product, an alternative bulking agent is required which not only has the physical properties of the nutritive sweetener being substituted, but also avoids particular metabolic problems and does not present any new ones.

It is probable that many consumers are totally unaware that the successful replacement of nutritive sweeteners in re-formulated foods depends to a very large extent on reproducing appropriate physical properties. In fact, it has been stated that 'even if sugar did not taste sweet, it would continue to be a major ingredient in many food products'.[1] Despite the food industry being in many respects conservative and traditional, many food products have been developed over the years that exploit the physical and chemical properties of nutritive sweeteners rather than their sweetness. Therefore, the replacement of nutritive sweeteners by non-nutritive sweeteners in foods is rarely a case of simple replacement of sweetness, but requires the innovative talents of product-development scientists in order to create satisfactory products.

A review of the properties of nutritive sweeteners, other than sweetness, responsible for their widespread use in foods is therefore necessary and should concentrate on sucrose since this is, despite the advent of high-fructose syrups, the major carbohydrate sweetener. The properties of the non-nutritive sweeteners permitted for use throughout the world will be discussed since these properties frequently restrict potential applications. Following this, an assessment will be given of the food-market categories that are ripe for exploiting the use of non-nutritive sweeteners, with examples of their successful incorporation into foods. Finally, the impact of future technical developments on non-nutritive-sweetener usage will be assessed since the increasing concern about obesity, diabetes and dental caries and their relation to diet will undoubtedly result in increased pressure on traditional nutritive-sweetener markets.

## 2. DEFINITION OF NON-NUTRITIVE SWEETENERS

For the purposes of this discussion the United States Food and Drug Administration's definition of non-nutritive sweeteners is appropriate. The Code of Federal Regulations[2] states that non-nutritive sweeteners are 'substances having less than 2 per cent of the caloric value of sucrose per equivalent unit of sweetening capacity'. Thus, even though the dipeptide sweetener aspartame (α-L-aspartyl-L-phenylalanine methyl ester) and the protein extracted from the West African plant *Thaumatococcus daniellii*, Talin Protein Sweetener®*, contribute calories they are classified as non-nutritive sweeteners because they are both more than 50 times sweeter than sucrose. On the other hand, if cyclamate, which has 30–50 times the sweetness of sucrose, was metabolised as carbohydrate or protein it would be classified as a nutritive sweetener because it would have marginally more than 2% of the caloric value of sucrose for each equivalent unit of sweetening capacity.

## 3. PROPERTIES OF NUTRITIVE SWEETENERS IN FOODS

In order to evaluate the technical potential of non-nutritive sweeteners in the major nutritive-sweetener sectors, it is first important to review the use of nutritive sweeteners in each food-market sector. It then becomes possible to project, on the basis of these technical reasons, what properties in addition to sweetness are required of nutritive sweeteners and hence to identify those markets in which a non-nutritive sweetener alone would be unable to find application. Since sucrose is the major nutritive sweetener in use, it will be referred to throughout the discussion.

### 3.1. Soft Drinks
In all categories of soft drinks, be they carbonated, still, or squash-concentrate, sucrose is used for its sweetness. It is frequently stated that sucrose provides mouthfeel in these applications, and while this is undoubtedly true it is an incidental factor not central to the use of sucrose. In

* Talin Protein Sweetener® is a registered trade mark of Tate & Lyle PLC, UK.

some cases the perception of mouthfeel or 'body' is closely correlated with the quality of sweetness. Thus saccharin, which is generally considered[3] to have a poor sweetness quality quite different from that of sucrose, does not contribute significantly to mouthfeel. However, the use of saccharin in conjunction with small quantities of flavour modifiers, e.g. lactose, calcium chloride or cream of tartar, results in products with significantly improved sweetness profiles closer to that of sucrose and such products have a perceived mouthfeel that is also much more sucrose-like.[4-8]

Water is the bulking agent in soft drinks and hence there is no technical barrier to replacing all of the nutritive sweeteners used in soft drinks with non-nutritive sweeteners of equivalent sweetness. That the non-nutritively sweetened proportion of the soft-drinks market hovers between 10 and 15 % is clear testimony of the sweetness quality of nutritive sweeteners. The sensory and technical deficiencies of currently approved non-nutritive sweeteners are perhaps overemphasised, while there may be a tendency to exaggerate the dietary concern of the average consumer.

### 3.2. Baked Products
Sucrose has a major influence on the structure of cakes, raising the temperature of coagulation of egg and delaying the gelation of starch.[9] Sucrose also affects protein and starch hydration, and the competition for water governs the physical properties of the baked product. Crystal size is important, particularly during the creaming of fat and sugar, which determines cake aeration. Sucrose is also an important humectant, i.e. it resists changes in moisture content and this is a factor governing the shelf-life of cakes.[10]

The sweetness of sucrose in baked goods *is* of importance, but its additional functional and bulking properties have such an important bearing on the overall eating quality of the finished product that it is difficult to imagine that any non-nutritive sweetener will find wide-spread application in this area—that is, unless used in combination with a suitable low-calorie bulking agent that can function in a manner equivalent to sucrose.

### 3.3. Sugar Confectionery
Sucrose provides sweetness and bulk in sugar confectionery. It has the ability to remain in solution even when the degree of supersaturation is high, resulting in the sugar glass expected in certain of these products. Textural control of grained goods (fudges and fondants) depends on the

ability of sucrose to form minute crystals ( $< 15 \mu$m) which produce a dense suspension in a saturated syrup in the presence of milk products and fat. Thus it is not possible to replace sucrose in these products by a non-nutritive sweetener unless a suitably versatile bulking agent is available.

## 3.4. Chocolate Confectionery

Chocolate is a suspension of ground cocoa solids and sucrose in cocoa butter. Sucrose acts as a cheap filler and its sweetness balances the extreme bitterness of the cocoa. It is also important to the texture of the finished product because of the small crystals that are formed with conching, so giving chocolate its characteristic 'snap'. For sweetness alone, sucrose is obviously replaceable by a non-nutritive sweetener, but the physical properties of sucrose are not so easily reproduced.

## 3.5. Dairy Products/Frozen Confectionery

Sucrose gives sweetness, bulk and textural attributes to ice-cream and it has an important influence on its freezing point.[11] With the advent of 'soft-scoop' ice-cream, corn syrups have found increasing usage at the expense of sucrose, but the use of non-nutritive sweeteners alone in ice-cream is unlikely because the textural and bulking properties are difficult to replace. Frozen confections are potential outlets for non-nutritive sweeteners when used in combination with nutritive sweeteners because the bulking agent in this case is principally water.

## 3.6. Desserts

The predominant reason for using nutritive sweeteners in desserts is to provide sweetness, although they have a preservative function in pie fillings. They contribute texture to whipped desserts and custards, as well as bulk to table jellies. However, the principal functional carbohydrate in these applications, apart from table jellies, is starch. Hence there is potential for using non-nutritive sweeteners in many of these applications.

## 3.7. Preserves

Sweetness, bulk, texture and preservative effects are contributed to preserves by sucrose, and since the high osmotic pressure it generates is a major factor in preventing microbial spoilage it is not possible to replace sucrose totally with a non-nutritive sweetener. Partial replacement is possible when coupled with appropriate adjustment of pectin type and content.

TABLE 1

TECHNICAL REASONS FOR USING SUCROSE IN FOOD-MARKET SECTORS

| Market sector | Technical reasons for using sucrose | | | | | |
|---|---|---|---|---|---|---|
| | Sweetness | Bulk | Texture | Humectancy | Freezing point depression | Osmotic effects |
| Soft drinks | * | | | | | |
| Baked products | * | * | * | * | | |
| Sugar confectionery | * | * | * | | | |
| Chocolate confectionery | * | * | * | | | |
| Dairy/frozen | * | * | * | | * | |
| Desserts | * | * | | | | |
| Preserves/spreads | * | | | | | * |
| Canned goods | * | | | | | * |
| Pickles/sauces | * | | | | | |

## 3.8. Canned Goods

Sucrose is used primarily for sweetness and as a balance to the natural acidity of fruits and vegetables. In sufficient concentration to form a 'heavy' syrup, sucrose can actually have an adverse effect on fruit owing to osmotic action. The cellular fluid, at a lower solids content than the surrounding syrup, has a tendency to migrate into the surrounding syrup thus reducing textural quality. The converse is true if non-nutritive sweeteners are used since water migrates into the fruit cells causing them to swell. Thus non-nutritive sweeteners are suitable for widespread use in these applications.

## 3.9. Pickles and Sauces

Sucrose is used in proportions of up to *circa* 25 % in pickles and sauces, its sweetening properties and balancing the natural acidity being the principal reasons for its use. Therefore non-nutritive sweeteners are satisfactory substitutes in these products.

A summary of the major food manufacturing market sectors, the role(s) of nutritive sweeteners in each application and the reasons for using nutritive sweeteners are given in Table 1, although effects such as fermentability, colour development, flavour enhancement, and antioxidant properties have not been considered.

Clearly, textural and bulking properties are almost as vital in nutritive sweeteners as sweetness is, and for the successful incorporation of non-nutritive sweeteners into food products these functions must be fulfilled by alternative bulk- and texture-conferring agents.

## 4.   NON-NUTRITIVE SWEETENERS

The non-nutritive sweeteners considered in this section—saccharin, cyclamate, aspartame, Acesulfame-K®*, stevioside, and Talin Protein Sweetener (Fig. 1)—are permitted food ingredients under different national statutory regulations, although they are not universally allowed.

The characteristics of an ideal non-nutritive sweetener are: (1) a sweetness quality and profile identical to that of sucrose; (2) sensory and chemical stability under the relevant food-processing and storage conditions; (3) compatibility with other food ingredients and stability toward other constituents naturally present or intentionally added to foods; (4) complete safety, shown as freedom from toxic, allergenic and

* Acesulfame-K® is a registered trade mark of Hoechst AG, West Germany.

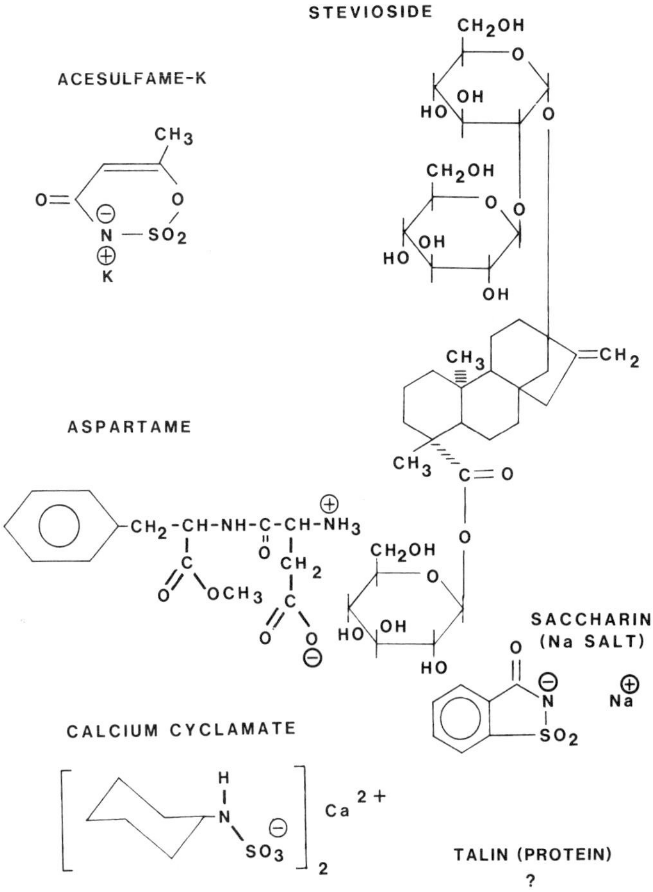

FIG. 1.   Structures of sweeteners under discussion.

other undesirable physiological properties; (5) complete freedom from metabolism in the body; and (6) high specific sweetness intensity. Obviously even the nutritive sweeteners currently used in foods would have difficulty meeting all of these criteria. However, the regulatory authorities in various parts of the world have been satisfied that the non-nutritive sweeteners they permit are safe for use in a designated range of food categories. Food-product-development scientists have therefore had to demonstrate their innovative skills in creating new food products sweetened with non-nutritive sweeteners to overcome whatever disadvantages each particular sweetener may have.

The main qualities of sweeteners depicted in Fig. 1 are given in Table 2.

TABLE 2

QUALITATIVE DESCRIPTION OF SELECTED NON-NUTRITIVE SWEETENERS

| Sweetener | Sweetness intensity[a] | Sweetness quality | Relative stability | Relative reactivity |
|---|---|---|---|---|
| Saccharin | 300 × | Sweet with a bitter metallic aftertaste | Very stable | Unreactive |
| Cyclamate | 30–50 × | 'Chemical' sweet, no aftertaste | Very stable | Unreactive |
| Aspartame | 180 × | Clean sweetness, sweet aftertaste | Unstable in liquid systems to low pH ($<4$) and elevated temperature | Potentially reactive |
| Acesulfame-K | 150 × | Sweet with bitter aftertaste, less pronounced than saccharin | Very stable | Probably unreactive |
| Stevioside | 300 × | Delayed onset, sweet with bitter aftertaste | Probably stable | Probably unreactive |
| Talin | 2 500 × | Slow onset of sweetness lingering liquorice aftertaste | Surprisingly stable | Potentially reactive |

[a] Sweetness intensity relative to sucrose (weight basis). Quoted intensities should be used only as a guide since precise values depend on application.

_learly, no single non-nutritive sweetener satisfies all the requirements of an ideal sweetener. Hence the use of these sweeteners is, of necessity, something of a compromise between the desire to produce low- or reduced-calorie or special-dietary foods and the need to ensure that newly created foods satisfy the stringent quality and consumer acceptability criteria that will dictate the success or failure of the new product.

### 4.1. Saccharin

Saccharin was first prepared in 1879 by Remsen and Fahlberg who reported[12] that 'the taste is perfectly pure'. However, it has been recognised for many years that saccharin has a distinct aftertaste[13] described as bitter or metallic, and it has been clearly demonstrated[14] that the aftertaste is an intrinsic characteristic of the saccharin molecule and not a property of impurities or decomposition products. Consumer responses to saccharin are frequently difficult to interpret. This has been explained in the belief that the ability to taste the bitterness of saccharin correlates with a genetically determined ability to perceive bitterness in compounds such as phenylthiocarbamide and n-propylthiouracil.[15] Thus some consumers are particularly sensitive to the bitter/metallic aftertaste of saccharin whereas others hardly perceive it at all. This has resulted in concerted efforts to formulate saccharin so that the bitter/metallic aftertaste is reduced or eliminated (for example, References 4–8). It is probably fair to say that although many of these formulations of saccharin (e.g. with lactose and cream of tartar, maltodextrin and calcium chloride, and aspartame) do improve the quality of sweetness none completely eliminates the saccharin aftertaste. This therefore remains an elusive goal but the rewards for success are great enough to ensure that efforts continue.

There are many literature reports giving the effects of subjecting

### TABLE 3
#### INFLUENCE OF TEMPERATURE AND pH ON THE STABILITY OF SACCHARIN

| pH | Decomposition after 1h (%) | | |
|----|--------|--------|--------|
|    | 100°C | 125°C | 150°C |
| 2·0 | 2·9 | 8·5 | 18·6 |
| 3·3 | 0 | 1·0 | 1·9 |
| 7·0 | 0·3 | 0·3 | 1·6 |
| 8·0 | 0 | 0 | 0 |

saccharin to extremes of temperature and pH.[16-19] DeGarmo ¢
describe the effects of heating saccharin solutions at pH values ranging
from 2 to 8 for 1 h at temperatures up to 150 °C (Table 3).

It is only at pH 2·0 that significant decomposition occurs—the
decomposition product being ammonium-o-sulphobenzoic acid. The
influence of water-soluble vitamins, essential amino-acids and olive oil on
the stability of saccharin at elevated temperatures has also been
studied.[18,19] Both vitamins and amino acids marginally increased the rate
of breakdown, but since the study was conducted with no pH regulation,
the increased decomposition was probably the result of deviations in pH
from the control. Olive oil appeared to have a protective effect (Table 4).

The net result of these studies is a clear demonstration of the stability of
saccharin under the normal range of conditions employed in food
processing and storage.

TABLE 4

INFLUENCE OF WATER-SOLUBLE VITAMINS, ESSENTIAL AMINO-ACIDS AND OLIVE OIL ON
THE STABILITY OF SACCHARIN

| Temperature (°C) | Decomposition of saccharin after 1 h (%) | | | |
|---|---|---|---|---|
| | Control | + Vitamins | + Amino acids | + Olive oil |
| 100 | 0 | 0 | 0 | 0 |
| 150 | 0 | 5 | 2 | 0 |
| 200 | 0 | 5 | 10 | 3 |
| 250 | 10 | 23 | 27 | 3 |

## 4.2. Cyclamate

Qualitatively, cyclamate is a satisfactory sweetener, although its flavour
has been described as being 'sweet-chemical'.[20] The major use of cyclamate
has been in admixture with saccharin at a cyclamate:saccharin weight-ratio
of 10:1. In this proportion, the cost-effectiveness and sweetness quality of
the mixture are optimised with cyclamate effectively eliminating the bitter
aftertaste of saccharin. Such mixtures are no longer permitted food-
additives in the UK or the USA, but there remain countries in which they
are permitted either for food use or as an ethical product (see Chapter 5).

Cyclamate is considered to be a chemically stable sweetener, although it
is less stable than saccharin.[18,19,21-4] Aqueous solutions of cyclamate are
hydrolysed to sulphuric acid and cyclohexylamine at very slow rates,
proportional to the hydrogen-ion concentration.[22] The half-life for a
solution of cyclamate at 125 °C buffered to pH 2·0 has been calculated to be

60 years.[21] The results of measurements of the stability at 30 and 44 °C and pH 2·1 are given in Fig. 2. Amino acids and water-soluble vitamins increase the decomposition rate of cyclamate at elevated temperature but, as with saccharin, this factor does not preclude the use of cyclamate in food formulations to any significant extent.

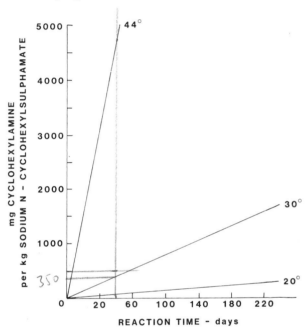

FIG. 2.    The stability of cyclamate at pH 2·1, and 20, 30 and 44 °C.

## 4.3. Aspartame

Aspartame is probably the non-nutritive sweetener closest to sucrose in flavour quality and profile. It has a clean sweetness with no associated bitter notes.[25] Any lingering sweet aftertaste which may be perceived is not unpleasant, although this does limit the thirst-quenching capabilities of aspartame-sweetened beverages.

The relative instability of aspartame is its major shortcoming, particularly in aqueous systems. Depending on the storage conditions and pH, aspartame undergoes internal condensation to give the corresponding diketopiperazine with the elimination of methanol, or is hydrolysed to the unesterified dipeptide (Fig. 3). The stability of aspartame in dry systems, such as beverage and dessert mixes, is reported to be very good, with 95 %

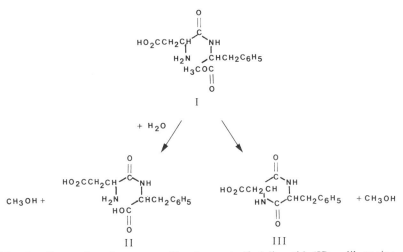

FIG. 3. Conversion of aspartame (**I**) to its unesterified dipeptide (**II**) or diketopiperazine (**III**).

recovery of aspartame after 3 years' storage under warehouse conditions, providing the product is packaged adequately.[26] However, aspartame is relatively unstable in aqueous environments, particularly at low pH (Fig. 4).

## 4.4. Acesulfame-K

There is a very limited published literature on Acesulfame-K. It has been claimed that the sweetness quality is not dissimilar to that of sucrose.[27] In

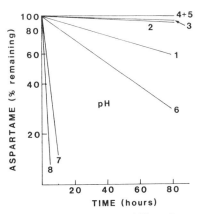

FIG. 4. Effect of pH and time on the stability of aspartame at 40 °C.

the author's opinion, this overstates the quality of Acesulfame-K since it has a clearly discernible bitter/metallic aftertaste. It is, however, superior to saccharin in that the bitter/metallic aftertaste is less prominent and this suggests that Acesulfame-K could find wide application in food systems. It is probably very much more costly to produce than saccharin, which is likely to restrict its market.

Acesulfame-K is a stable sweetener, particularly over the range of temperatures and pH values that foods are subjected to. Hydrolysis takes place only under extreme conditions[28] producing acetone, carbon dioxide, and ammonium sulphate ions. Aqueous solutions may be sterilised at pH 4 and 120 °C with no detectable decomposition. There is no detectable hydrolysis on subsequent storage of the sterilised solution at 40 °C for one month.

### 4.5. Stevioside

Stevioside is a diterpene glycoside extracted from the leaves of *Stevia rebaudiana*, the sweet herb of Paraguay, and is used extensively in Japan. Stevioside has a bitter aftertaste and a clear lingering sweetness,[29] although new structural variants are reported to have a cleaner sweetness quality (see Chapter 5, Section 6). Stevioside is sufficiently stable for incorporation into all types of food systems, although it is liable to enzymic hydrolysis, yielding its aglycone, steviol, which has shown evidence of anti-androgenic activity.[30]

### 4.6. Talin Protein Sweetener

Talin Protein Sweetener is the proprietary name given to the purified sweet extract obtained from the fruit of the West African plant *Thaumatococcus daniellii*. The sweetness profile is characterised by a delay in perception with a lingering sweet liquorice aftertaste[31] which limits its usefulness as the sole sweetener in a food system. It is currently permitted for use as a food ingredient in Japan where it is used widely, both in low-calorie foods and drinks and also as a flavour modifier at sub-sweetness threshold levels of addition.

Talin sweetener, a protein, is unexpectedly stable to extremes of pH and temperature. Stability is enhanced at lower pH values, allowing heat treatment at 100 °C for several hours without loss of sweetness at pH values less than 5·5. This renders it ideal for use in soft drinks (typically pH 2·8–3·5) which can be pasteurised or even subjected to UHT sterilisation.[32]

## 5.   NON-NUTRITIVE SWEETENERS IN FOOD SYSTEMS

The factors governing consumer acceptability of standard foods and beverages apply equally to the acceptance of low-calorie products and other special-dietary foods. Psychologically, it is important that the consumer considers a low-calorie or special-dietary version to be of equal quality to the equivalent product sweetened with nutritive sweeteners and will enjoy eating the product for its own sake. Diabetics and dieters will resist consuming special-dietary foods that are viewed as being of inferior quality. Their expectation is for products manufactured with non-nutritive sweeteners to be as acceptable as those made with nutritive sweeteners. It is this expectation which emphasises the need for food manufacturers to approach the development of new foods sweetened with non-nutritive sweeteners in a manner analogous to the development of nutritively sweetened foods.

As was pointed out earlier in this chapter, it is not sufficient for food manufacturers to develop special-dietary foods and beverages in which the nutritive sweetener has simply been replaced by a non-nutritive sweetener. In general, a completely new formulation is required in which the development goal is more likely to be equal acceptability of the non-nutritively sweetened version, rather than an exact duplication of the nutritively sweetened equivalent.

Until recently, it was not possible to produce baked goods and confectionery items sweetened with non-nutritive sweeteners that had acceptable sensory properties. With the recent regulatory approval[33,34] of the low-calorie bulking agent Polydextrose®*, food manufacturers now have the opportunity to create reduced-calorie baked goods and confectionery products that are claimed[35] to have the equivalent texture of the full-calorie equivalents. The very recent introduction of Polydextrose means, however, that there are no independent evaluations reported in the literature, so it is not possible to review critically the use of non-nutritive sweeteners in concert with Polydextrose at this time. It is probably fair to say, however, that the availability of low-calorie bulking agents will serve to increase the use of non-nutritive sweeteners only in certain market sectors.

The use of non-nutritive sweeteners in other applications has been extensively studied and there now follows a discussion of the key factors which must be considered when using non-nutritive sweeteners in each particular market sector.

---

* Polydextrose® is a registered trade mark of Pfizer Inc., USA.

## 5.1. Beverages

### 5.1.1. Carbonated Beverages

In addition to matching the sweetness provided by nutritive sweeteners, non-nutritive sweeteners in carbonated beverages need to be carefully selected for flavour, with particular attention to carbonation levels. Flavours ideally should be of high quality and free from undesirable notes that would be magnified in, for example, a saccharin-sweetened beverage. Carbonation level influences flavour and mouthfeel significantly, and this influence is even more pronounced in non-nutritively-sweetened beverages. As a general rule, with low-calorie carbonated beverages the greater the degree of carbonation (up to five volumes), the greater the acceptability.

Saccharin remains the major sweetener in low-calorie beverages which invariably suffer from its poor sweetness quality and bitter aftertaste. Aspartame is permitted in carbonated beverages in Canada, but careful stock control is crucial if its instability is not to result in extreme sweetness variability of products.

### 5.1.2. Dry Mixes

Both sugar-sweetened and non-nutritively-sweetened dry mixes are available commercially. They consist essentially of sweetener, acidulant, flavour, colour, and, occasionally, added filler. The important factors to consider in formulating reduced-calorie versions are quality of flavour, choice of acidulant, and the use of bulking agents to confer mouthfeel. Malic, citric and fumaric acids are available as acidulants, and the choice is principally an economic one although fumaric acid has the advantage of being non-hygroscopic and hence assists storage stability particularly in dietary formulations. Bulking agents in low-calorie formulations can also reduce the importance of expensive packaging by assisting with moisture control. Maltodextrin is the preferred choice of recent years, principally because of economic factors. Aspartame is an ideal sweetener for this type of application because stability is not a crucial problem. With the recent re-approval of aspartame in the USA and elsewhere in the world, new products are already appearing on the market.

## 5.2. Baked Goods

The role of sucrose in baked goods was discussed earlier in this chapter, where it was pointed out that sucrose is crucial to the generation of a satisfactory texture. When non-nutritive sweeteners replace sucrose in baked goods, the appearance, volume and texture of the finished product tend to be unacceptable.[36] The use of bulking agents such as Polydextrose may open this market sector to the use of non-nutritive sweeteners to an extent yet to be determined.

## 5.3. Confectionery

It is difficult to imagine that non-nutritive sweeteners will find significant usage in chocolate confectionery, not only on the grounds of legislative constraints but also because of the bulking and textural properties of sucrose. In sugar confectionery, non-nutritive sweeteners may find more use because of the advent of materials such as Polydextrose.

## 5.4. Dairy Products/Frozen Desserts

A typical food product in this category is ice-cream. Normal ice-cream formulations contain approximately 12% fat and provide about 210 kcals $100 g^{-1}$. In order to create a satisfactory low-calorie product, the obvious ingredients to reduce are fat and carbohydrate. Reductions in the fat content require substantial modifications to the formulation in order to achieve a satisfactory texture. This is achieved by careful selection of emulsifiers and stabilisers.[36] Emulsifiers, such as mono- and di-glycerides of edible fats, increase creaminess and smoothness whereas stabilisers, such as gelatin, and guar, carrageenan and cellulose gums, bind water and prevent large crystal growth. Formulations have been developed with both reduced fat and carbohydrate—the carbohydrate is replaced by a non-nutritive sweetener such as saccharin (Table 5). There is little doubt, however, that a product so formulated will be inferior in quality to full fat/carbohydrate equivalents and hence the market potential is small.[37] Aspartame may also be formulated in an ice-cream equivalent[25] with a 50% reduction in caloric value.

It is also claimed[25] that flavouring plain yoghurt with a dry mix of sweetener, flavour and colour is feasible. The sweetener in question is

TABLE 5
COMPOSITION OF DIABETIC ICE-CREAMS

|  | Percentage |
| --- | --- |
| Butterfat | 14 |
| Milk solids | 9 |
| Sorbitol solution (83·7% solids) | 15 |
| Stabiliser | 0·2 |
| Saccharin | 0·02 |
| Total solids | 38·22% |

aspartame, and, since the caloric addition to plain yoghurt is very small, a calorie reduction of 50 % is attainable. It is admitted, however, that this example is merely an indication of the diversity of opportunity open to product development.

## 5.5. Dessert Mixes

Gelatin desserts sweetened with non-nutritive sweeteners have been formulated successfully but texture can be a serious problem. Sorbitol or mannitol may be used for diabetic products, primarily as diluent dispersants, fillers and bodying agents.[36] Such sugar alcohols are obviously less successful ingredients for low-calorie products where different Bloom-strength gelatins or alternative gelling agents, e.g. carrageenan and alginate, are necessary to create a satisfactory product.[38] Aspartame-sweetened gelatin desserts have been created successfully[26] but improved products are obtained when only a portion of the sucrose has been substituted. Puddings are normally set with starches or alginates but similar considerations apply as for gelatin desserts.[39] Calorie reductions attainable are generally lower than for gelatin desserts because of the fixed calorie contribution from the milk in these products[25] (Table 6).

## 5.6. Preserves

Normal jams contain in excess of 65 % soluble solids, which ensures that there is insufficient available water for microbial growth. Low-calorie jams

TABLE 6

CHOCOLATE   PUDDING   OR   PIE   FILLING   SWEETENED   WITH
ASPARTAME[25a]

| Ingredient | Composition of dry mix (%) |
|---|---|
| Starch thickeners | 68·0 |
| Cocoa | 12·9 |
| Alginate | 6·5 |
| Trisodium phosphate | 5·2 |
| Aspartame | 2·6 |
| Sodium dihydrogen phosphate | 2·3 |
| Calcium acetate | 1·9 |
| Sodium chloride | 0·6 |
| Flavour and colour | Minimal |
|  | 100·0 |

[a] Calorie reduction compared to sugar equivalent equals 43 %.

can be prepared and may have a solids content as low as 15–20%.[36] At these levels it is necessary to modify the choice of pectin from one of relatively high methoxyl content to one of low methoxyl content in order to obtain products of sufficient gel strength. It then becomes feasible to incorporate a non-nutritive sweetener, such as saccharin, to replace the sweetness lost in reducing the sucrose content. Stability of sweetener is crucial and for this reason it is not possible to use aspartame in low-calorie jams.

### 5.7. Canned Products

In canned fruits, it is relatively straightforward to replace sucrose or corn syrup with a non-nutritive sweetener. The amount of non-nutritive sweetener obviously depends on the natural sweetness of the fruit and the sweetness level desired. If a greater viscosity is required, this can be achieved by judicious use of pectin. As with preserves, the stability of the sweetener is important and hence the use of aspartame in this application is precluded. Saccharin, cyclamate, Acesulfame-K and stevioside should all prove acceptable.[40]

### 5.8. Breakfast Cereals

Pre-sweetened breakfast cereals are almost invariably consumed by children whose need for special-dietary or low-calorie foods is clearly restricted. However, patents have been published that describe the application of aspartame as a sweetener in breakfast cereals.[41,42] One claimed[25] advantage of sweetening with a non-nutritive sweetener is that the nutrient density of the product is increased, although realistically this argument has little merit.

### 5.9. Non-Food Uses

Pharmaceutical preparations frequently utilise sweetness in order to improve palatability.[43] In the majority of cases, sucrose is the sweetener of choice, and it is, in fact, one of the better masking agents in this type of application.[44] However, pharmaceutical preparations sweetened with sucrose may not be suitable for diabetics or for those on strict calorie-controlled diets. Consequently, the use of non-nutritive sweeteners in liquid pharmaceuticals has received much attention.[44–6] In tablets, the choice of non-nutritive sweetener is dictated first by sensory properties and second by cost. Stability is not a factor, but compatibility with active ingredients is. In liquid preparations, shelf stability is important and the use of saccharin or saccharin/cyclamate mixtures has been preferred. The use of Talin Protein

Sweetener has also been proposed for this type of application because its unique sensory profile is ideally suited to mask the lingering bitter aftertaste of many drugs.[32]

## 6. FUTURE DEVELOPMENTS

The two key developments that will have a major impact on the use of non-nutritive sweeteners in food systems are:

1.   The development and regulatory approval of a non-nutritive sweetener that combines the attributes of sucrose-like quality of sweetness with sufficient stability to heat and acidity for it to be used in all food and beverage applications; and
2.   The development and regulatory approval of a low-calorie bulking agent capable of reproducing the physical properties of sucrose at an equivalent cost, or preferably substantially below the cost of sucrose per unit weight.

The development of a sucrose-quality sweetener with high stability to heat and low pH theoretically would open up most food-market sectors for penetration, and if such a sweetener was to become available along with a cheap low-calorie bulking agent this would have a major impact on the markets for low-calorie and special-dietary foods. There is little doubt that a number of food companies are actively pursuing these goals although any major developments prior to the 1990s seem unlikely in view of the long time-scale needed for obtaining regulatory approval.

## 7. CONCLUSIONS

Food-industry product-development personnel undoubtedly have the ability to create a wide range of acceptable low-calorie foods and beverages. The food industry has proved to be capable of overcoming or reducing the problems associated with the non-nutritive sweeteners currently approved throughout the world, but shortcomings such as the bitterness of saccharin and the instability of aspartame are intrinsic properties of the sweeteners which still limit their application. Hence products sweetened with saccharin and aspartame have achieved only a limited penetration of the market. The development of a low-cost bulking agent would clearly open-up further markets to non-nutritive sweeteners that were hitherto unavailable, and therefore would have a major impact on the use of non-nutritive sweeteners in food systems.

## REFERENCES

1. MACKAY, D. A. M. and VINCENT, P. M. (1978). In: *Health and Sugar Substitutes*, Proceedings ERGOB Conference, B. Guggenheim (Ed.), Karger, Basel, 321.
2. US CODE OF FEDERAL REGULATIONS (1979). 21. *Food and Drugs*, Section 170.6, Part 19.
3. SALANT, A. (1975). In: *Handbook of Food Additives*, 2nd edn. T. E. Furia (Ed.), CRC Press, Ohio, 533.
4. LIGGETT, J. J. and HOERRES, W. E. (1972). US Patent 3 684 529.
5. BLIZNAK, J. B. (1973). US Patent 3 773 526.
6. EISENSTADT, M. E. (1973). US Patent 3 743 518.
7. HILL, J. A. and LAVIA, A. L. (1972). US Patent 3 695 898.
8. LAVIA, A. L., O'LAUGHLIN, R. L. and WALTON, R. W. (1972). US Patent 3 704 138.
9. NICOL, W. M. (1979). In: *Sugar: Science and Technology*, G. G. Birch and K. J. Parker (Eds.), Applied Science Publishers Ltd, London, 211.
10. SEILER, D. A. L. (1969). BFMIRA Symposium NO. 4, *Relative Humidity in the Food Industry*, 28.
11. LACHMANN, A. (1972). *The Role of Sucrose in Foods*, International Sugar Research Foundation, Inc., Bethesda, Maryland, 237.
12. REMSEN, I. and FAHLBERG, C. (1879). *Amer. Chem. J.*, **1**, 426.
13. STUTZEN, A. (1886). *Biedermanns Zent.*, **15**, 64.
14. HELGREN, F. J., LYNCH, M. J. and KIRCHMEYER, F. J. (1955). *J. Amer. Pharm., Assoc.*, **44**, 353.
15. BARTOSHUK, L. M. (1979). *Science*, **205**, 934.
16. DEGARMO, O., ASHWORTH, G. W., EAKER, C. M. and MUNCH, R. H. (1952). *J. Amer. Pharm. Assoc.*, **41**, 17.
17. VON TANFEL, K. and NATON, J. (1926). *Zeit. angewandte Chemie*, **39**, 224.
18. KROYER, G. and WASHUTTL, J. (1979). *Z. Ernahrungswiss*, **18**, 139.
19. KROYER, G. and WASHUTTL, J. (1979). *Lebensm.-Wiss. u.-Technol.*, **12**, 284.
20. LARSON, N. (1975). *Sensory Properties of Flavoured Gelatins Containing Sucrose or Synthetic Sweeteners*, MS Thesis, University of California, Davis.
21. TALMAGE, J. M., CHAFETZ, L. and ELEFANT, M. (1968). *J. Pharm. Sci.*, **57**, 1073.
22. SWALLOW, W. H. (1975). *New Zealand J. Sci.*, **18**, 541.
23. HRDLICKA, J. and JANICEK, G. (1971). *Z. Lebensm. Forsch.*, **145**, 291.
24. MARUYAMA, K., KAWANABE, K., FUJIOKA, A. and IIJIMA, S. (1971). *J. Pharm. Soc. Japan*, **91**, 579.
25. BECK, C. I. (1977). In: *Low Calorie and Special Dietary Foods*, B. K. Dwivedi (Ed.), CRC Press, Florida, 59.
26. SCOTT, D. (Ed.), (1974). *Proc. IV Internat. Congr. Fd. Sci. Technol.*, **1**, 468.

27. CLAUSS, K., LÜCK, E. and VON R. LIPINSKI, G.-W. (1976). Z. Lebensm. Unters.-Forsch., **162**, 37.
28. ARPE, H.-J. (1978). In: Health and Sugar Substitutes, Proceedings ERGOB Conference, B. Guggenheim (Ed.), Karger, Basel, 178.
29. POMARET, M. and LAVIEILLE, R. (1931). Bull. Soc. Chim. Biol., **13**, 1248.
30. DORFMAN, R. I. and NES, W. R. (1960). Endocrinology, **67**, 282.
31. HIGGINBOTHAM, J. D., LINDLEY, M. G. and STEPHENS, J. P. (1981). In: The Quality of Foods and Beverages, G. Charalambous and G. Inglett (Eds.), Academic Press, New York, 91.
32. HIGGINBOTHAM, J. D. (1979). In: Developments in Sweeteners—1, C. A. M. Hough, K. J. Parker and A. J. Vlitos (Eds.), Applied Science Publishers Ltd, London, 87.
33. Food Additives and Contaminants Committee Review of Remaining Classes of Food Additives used as Ingredients in Food: Report on the Review of Bulking Aids (1980). HMSO, London.
34. ANON. (1981). Food Chem. News, June 8, 45.
35. ANON. (1981). Food Processing, July, 36.
36. SALANT, A. (1975). In: Handbook of Food Additives, T. E. Furia (Ed.), CRC Press, Ohio, 523.
37. ROBERTS, C. G. and DAHLE, C. D. (1954). Ice Cream Trade J., April, 66.
38. ZABICK, M. E., MILLER, G. A. and ALDRICH, P. A. (1962). Food Tech., **16**, 87.
39. HORN, L. J. (1971). US Patent 3 563 769.
40. ANDERSON, E. E., ESSELEN, W. B. and FELLERS, C. R. (1953). J. Amer. Diet. Assoc., **29**, 770.
41. BAGGERLEY, P. A. (1975). US Patent 3 955 000.
42. ROUSSEAU, P. M. (1976). US Patent 3 947 600.
43. BROOKS, L. G. (1965). Chem. Drug., 421.
44. SHORT, G. R. A. (1960). Pharm. J., **131**, 565.
45. LYNCH, M. J. and GROSS, H. M. (1960). Drug Cosmet. Ind., **87**, 324.
46. ENDICOTT, C. J. and GROSS, H. M. (1959). Drug Cosmet. Ind., **85**, 176.

# INDEX